ARCHITECTURAL DESIGN

GUEST-EDITED BY
FABIO GRAMAZIO AND
MATTHIAS KOHLER

MADE BY ROBOTS
CHALLENGING ARCHITECTURE AT A LARGER SCALE

03 / 2014

ARCHITECTURAL DESIGN
MAY/JUNE 2014
ISSN 0003-8504

PROFILE NO 229
ISBN 978-1118-535486

IN THIS ISSUE

ARCHITECTURAL DESIGN

GUEST-EDITED BY
FABIO GRAMAZIO AND
MATTHIAS KOHLER

MADE BY ROBOTS: CHALLENGING ARCHITECTURE AT A LARGER SCALE

5 EDITORIAL
   Helen Castle

6 ABOUT THE GUEST-EDITORS
   Fabio Gramazio and Matthias Kohler

8 SPOTLIGHT
   Visual highlights of the issue

14 INTRODUCTION
   Authoring Robotic Processes
   Fabio Gramazio, Matthias Kohler and Jan Willmann

22 Integrating Robotic Fabrication in the Design Process
   Michael Budig, Jason Lim and Raffael Petrovic

44 Mesh-Mould: Robotically Fabricated Spatial Meshes as Reinforced Concrete Formwork
   Norman Hack and Willi Viktor Lauer

54 Robots and Architecture: Experiments, Fiction, Epistemology
   Antoine Picon

60 Entrepreneurship in Architectural Robotics: The Simultaneity of Craft, Economics and Design
   Jelle Feringa

   66 Odico Formwork Robotics
      Asbjørn Søndergaard

   68 RoboFold and Robots.IO
      Gregory Epps

   70 Machineous
      Andreas Froech

   72 ROB Technologies
      Tobias Bonwetsch and Ralph Bärtschi

   74 GREYSHED
      Ryan Luke Johns

EDITORIAL BOARD
Will Alsop
Denise Bratton
Paul Brislin
Mark Burry
André Chaszar
Nigel Coates
Peter Cook
Teddy Cruz
Max Fordham
Massimiliano Fuksas
Edwin Heathcote
Michael Hensel
Anthony Hunt
Charles Jencks
Bob Maxwell
Jayne Merkel
Peter Murray
Mark Robbins
Deborah Saunt
Patrik Schumacher
Neil Spiller
Leon van Schaik
Michael Weinstock
Ken Yeang
Alejandro Zaera-Polo

76 Computation or Revolution
Philippe Morel

88 Changing Building Sites: Industrialisation and Automation of the Building Process
Thomas Bock and Silke Langenberg

100 In-Situ Fabrication: Mobile Robotic Units on Construction Sites
Volker Helm

108 Towards Robotic Swarm Printing
Neri Oxman, Jorge Duro-Royo, Steven Keating, Ben Peters, Elizabeth Tsai

116 Machines for Rent: Experiments by New-Territories
François Roche and Camille Lacadée

126 COUNTERPOINT
Crisis! What Crisis? Retooling for Mass Markets in the 21st Century
Tom Verebes

134 CONTRIBUTORS

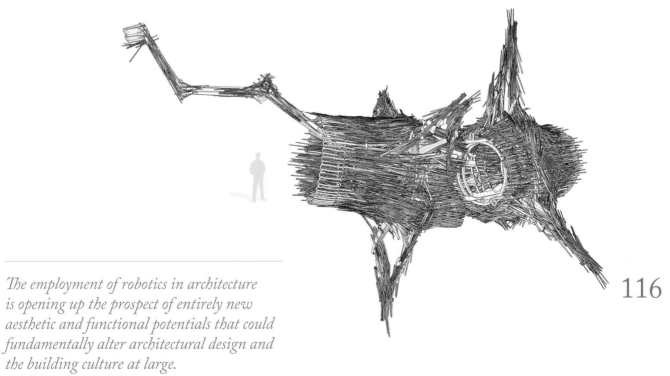

The employment of robotics in architecture is opening up the prospect of entirely new aesthetic and functional potentials that could fundamentally alter architectural design and the building culture at large.
— *Gramazio & Kohler*

ARCHITECTURAL DESIGN
MAY/JUNE 2014
PROFILE NO 229

Editorial Offices
John Wiley & Sons
25 John Street
London WC1N 2BS
UK

T: +44 (0)20 8326 3800

Editor
Helen Castle

Managing Editor (Freelance)
Caroline Ellerby

Production Editor
Elizabeth Gongde

Prepress
Artmedia, London

Art Direction and Design
CHK Design:
Christian Küsters
Sophie Troppmair

Printed in Italy by Printer Trento Srl

All Rights Reserved. No part of this publication may be reproduced, stored in a retrieval system or transmitted in any form or by any means, electronic, mechanical, photocopying, recording, scanning or otherwise, except under the terms of the Copyright, Designs and Patents Act 1988 or under the terms of a licence issued by the Copyright Licensing Agency Ltd, 90 Tottenham Court Road, London W1T 4LP, UK, without the permission in writing of the Publisher.

Subscribe to ⌭

⌭ is published bimonthly and is available to purchase on both a subscription basis and as individual volumes at the following prices.

Prices
Individual copies: £24.99 / US$45
Individual issues on ⌭ App
for iPad: £9.99 / US$13.99
Mailing fees for print may apply

Annual Subscription Rates
Student: £75 / US$117 print only
Personal: £120 / US$189 print and iPad access
Institutional: £212 / US$398 print or online
Institutional: £244 / US$457 combined print and online
6-issue subscription on ⌭ App
for iPad: £44.99 / US$64.99

Subscription Offices UK
John Wiley & Sons Ltd
Journals Administration Department
1 Oldlands Way, Bognor Regis
West Sussex, PO22 9SA, UK
T: +44 (0)1243 843 272
F: +44 (0)1243 843 232
E: cs-journals@wiley.com

Print ISSN: 0003-8504
Online ISSN: 1554-2769

Prices are for six issues and include postage and handling charges. Individual-rate subscriptions must be paid by personal cheque or credit card. Individual-rate subscriptions may not be resold or used as library copies.

All prices are subject to change without notice.

Rights and Permissions
Requests to the Publisher should be addressed to:
Permissions Department
John Wiley & Sons Ltd
The Atrium
Southern Gate
Chichester
West Sussex PO19 8SQ
UK

F: +44 (0)1243 770 620
E: permreq@wiley.com

Front cover: Pascal Genhart and Tobias Wullschleger, Nested Voids, Design of Robotic Fabricated High Rises, Architecture and Digital Fabrication, Future Cities Laboratory (FCL), Singapore-ETH Centre for Global Environmental Sustainability (SEC), 2012. © Gramazio & Kohler, Architecture and Digital Fabrication, ETH Zurich

Inside front cover: URStudio controlling a tiling process. © ROB Technologies AG

03 / 2014

EDITORIAL
Helen Castle

Over the last decade, the names of Fabio Gramazio and Matthias Kohler have become synonymous with robotics in architecture. At ETH Zurich, where they share the Chair in Architecture and Digital Fabrication, they became in 2005 the first multipurpose fabrication laboratory in the discipline of architecture to employ an industrial robot. As a result, they have almost single-handedly driven robotic research in the field.

Robotics over the next few years will without a doubt become a game changer for the entire construction industry. Greater mechanisation both on- and off-site will enable manual labour to be minimised, as a means of achieving greater efficiencies and cost savings. This makes it a crucial period of transition for architecture. How can designers take early ownership of this space in order to ensure their position in the field and concentrate technological efforts towards high-quality building design and construction rather than just solely on operational productivity? The work of Gramazio and Kohler at ETH Zurich and in practice keeps the focus almost wholly on the value of robotics for the discipline of architecture. Their research highlights such key questions as: How can robotics expand the range of production and design options for architects by increasing the potential for greater material differentiation and complexity of form? What are the possibilities for applying robotics at the large scale? How might, for instance, in large-scale constructions such as the high-rise, the use of the robot cause a shift away from standard parts to the bespoke?

As Gramazio and Kohler point out in their introduction, robotics in architecture has the potential to recast the entire field as a practice: 'the modern division between intellectual work and manual production, between design and realisation, is being rendered obsolete' (p 14). Nowhere is it more apparent in this issue that both the approach and profession of architects are being reshaped than in the exciting section on entrepreneurship in architectural robotics (pp 60–75). Startups are proving an entirely new and innovative nexus between academic research and the construction industry. On a tech model, agile outfits are stepping in and attempting to benefit from current gaps in knowledge and technologies. Introduced by Jelle Feringa, Chief Technology Officer at Odico Formwork Robotics, a 15-strong company that specialises in formwork production and research and development (R&D) technology, this extended article features five startups located in Funen in Denmark, London, Los Angeles, Princeton in New Jersey and Zurich.

The extent to which robotics in architecture has the potential to bring about real change globally and influence the quality of the wider built environment is brought to the fore in Tom Verebes' Counterpoint to the issue (pp 126–33). Nowhere are the stakes higher than in China and Southeast Asia where rapid urbanisation has meant that there is a relentless and almost unmeetable demand for high-rise housing in burgeoning cities. From his base in Hong Kong, Verebes questions the real possibilities for mass adoption of robotics in construction in Chinese cities in a way that could lead to the development of differentiated and bespoke architectures. What is apparent, though, is that it is only with the type of pioneering research that Gramazio and Kohler and their colleagues are undertaking at ETH Zurich and the associated Future Cities Laboratory (FCL) in Singapore that architects will ever have the hope of being significant players at the table when the international construction industry comes to the point of rolling out on-site automated building techniques. ᗪ

Text © 2014 John Wiley & Sons Ltd. Image © Illustration by Frances Castle

Gramazio & Kohler in cooperation with Bearth & Deplazes, Gantenbein vineyard facade, Fläsch, Switzerland, 2006
top: The masonry bond with the gaps between the bricks creates subdued interior lighting. The sunlight shining through produces fascinating illumination effects.

Gramazio & Kohler and Raffaello D'Andrea in cooperation with ETH Zurich, *Flight Assembled Architecture*, FRAC Centre, Orléans, France, 2012
bottom left: The final installation of *Flight Assembled Architecture* consists of over 1,500 modules which are placed by a multitude of quad-rotor helicopters, collaborating according to mathematical algorithms that translate digital design data to the behaviour of the flying machines.

Fabio Gramazio and Matthias Kohler, *Structural Oscillations*, Architecture and Digital Fabrication, ETH Zurich, Venice Architecture Biennale, 2008
bottom right: Interior view of the Swiss Pavilion with its 100-metre (328-foot) long robotic fabricated wall.

ABOUT THE GUEST-EDITORS
FABIO GRAMAZIO AND MATTHIAS KOHLER

Fabio Gramazio and Matthias Kohler are architects with multidisciplinary interests ranging from computational design and robotic control and fabrication, to material innovation. In 2000 they founded the architecture practice Gramazio & Kohler, which has realised numerous award-winning projects integrating novel architectural designs within a contemporary building culture. Current projects include the design of the NEST research platform, a future living and working laboratory for sustainable building construction. Built work ranges from international exhibitions and private and public buildings, to large-scale urban interventions, including the Gantenbein vineyard facade (Fläsch, Switzerland, 2006), the Tanzhaus theatre for contemporary dance (Zurich, 2007), the Christmas lights for Bahnhofstrasse, Zurich (2003–5), and the sWISH* pavilion at the Swiss National Exposition Expo.02 (Biel, Berne, 2002).

Opening also the world's first architectural robotic laboratory at ETH Zurich, Gramazio & Kohler's research has been formative in the field of digital architecture, setting precedence and de facto creating a new research field merging advanced architectural design and additive fabrication processes through the customised use of industrial robots. This ranges from 1:1 prototype installations to robotic fabrication at a large scale, which is explored at the Future Cities Laboratory of the Singapore ETH-Centre for Environmental Sustainability (SEC) and exclusively featured in this issue of △.

Gramazio & Kohler were awarded the Swiss Art Award in 2004, the Acadia Award for Emerging Digital Practice in 2009, and the Global Holcim Innovation Prize in 2012. Their innovative explorations have contributed to numerous exhibitions around the world, with installations such as *Structural Oscillations* at the 2008 Venice Architecture Biennale, *Pike Loop* at the Storefront for Art and Architecture in New York in 2009, and *Flight Assembled Architecture* at the FRAC Centre, Orléans, France, in 2011–12. Their work has been published in many journals, books and media, and was first documented in their book *Digital Materiality in Architecture* (Lars Müller Publishers, 2008). Their recent research is outlined and theoretically framed in the book *The Robotic Touch: How Robots Change Architecture* (Park Books, 2014). Together with leading researchers in architecture, material sciences, computation and robotics, they have just launched the first architectural National Centre of Competence in Research on Digital Fabrication. △

Text © 2014 John Wiley & Sons Ltd. Images: p 6(t) © Ralph Feiner; p 6(bl) © FRAC Centre, photography François Lauginie; p 6(br) © Alessandra Bello; p 7 © Juventino Mateo

SPOTLIGHT

Gramazio & Kohler and Raffaello D'Andrea in cooperation with ETH Zurich

Flight Assembled Architecture, FRAC Centre, Orléans, France, 2011
Leaving the common workspace of conventional digitally controlled production machines, flying robots can operate freely in airspace. Here they assemble collaboratively over 1,500 building modules into a vertical urban structure.

RoboFold

Sartorial Techtonics folded panel system, 2013
RoboFold developed a folded panel system based on textile folding patterns for Andrew Saunders of Rensselaer Polytechnic Institute (RPI) in New York. The project takes its inspiration from traditional pleating techniques, and specifically the box pleat. Folds in the cloth are translated directly to metal through material simulation and physical experimentation. A total of 11 panels were produced as a part of 1:1 facade mock-up, and shipped to the US to be assembled and exhibited at RPI.

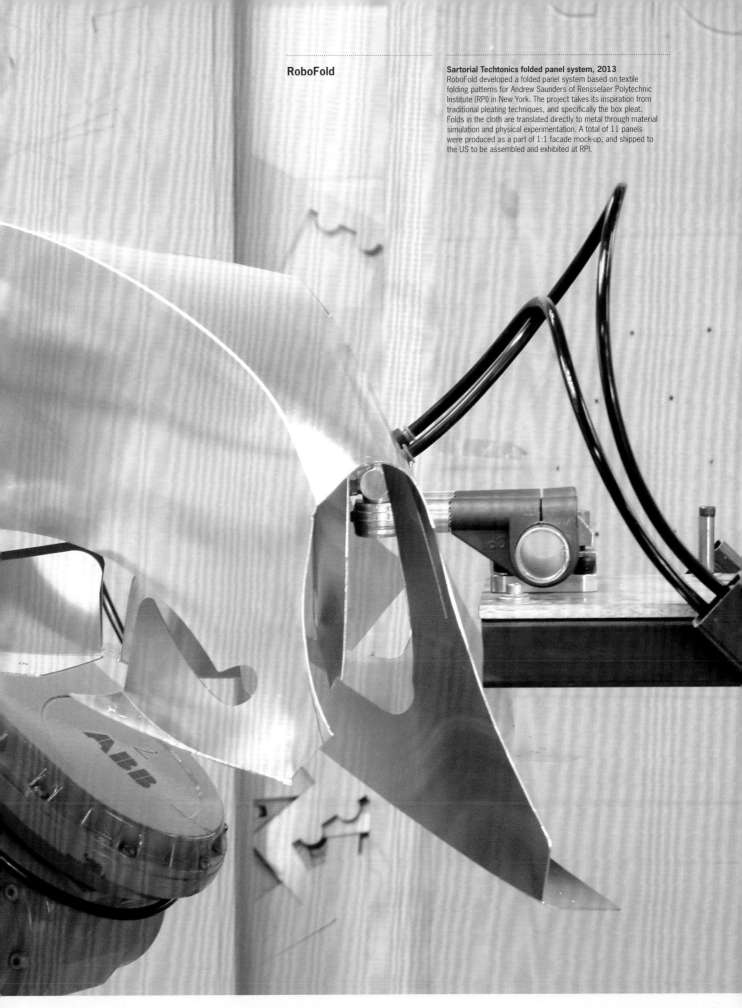

Original patent of the first industrial robot

Programmed Article Transfer, by George Charles Devol, Jr., issued **13 June 1961**
Devol applied for the patent on 10 December 1954, the document extending over a mere three pages.

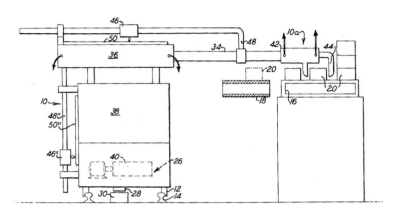

June 13, 1961   G. C. DEVOL, JR   2,988,237
PROGRAMMED ARTICLE TRANSFER

Filed Dec. 10, 1954   3 Sheets-Sheet 1

Fig.1

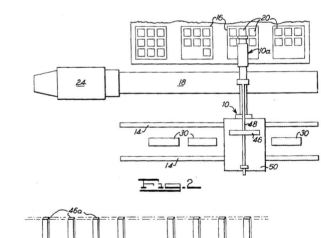
Fig.2

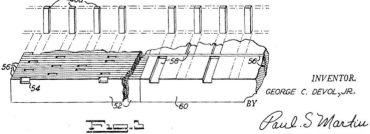
Fig.3

INVENTOR.
GEORGE C. DEVOL, JR.
BY Paul S Martin
ATTORNEY

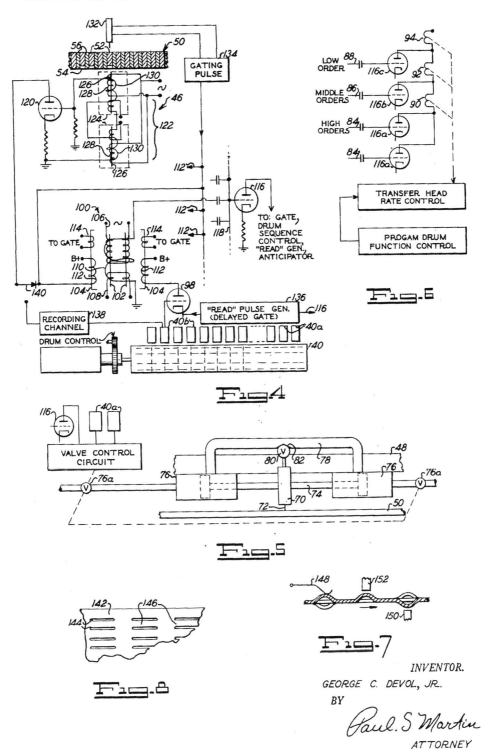

INTRODUCTION
Fabio Gramazio, Matthias Kohler and Jan Willmann

AUTHORING ROBOTIC PROCESSES

Over the past decade, robotic fabrication in architecture has succeeded where early digital architecture failed: in the synthesis of the immaterial logic of computers and the material reality of architecture where the direct reciprocity of digital designs and full-scale architectural production is enabled. With robots, it is now possible to radically enrich the physical nature of architecture, to 'inform' material processes and to amalgamate computational design and constructive realisation as a hallmark feature of architecture in the digital age, leading to the emergence of a phenomenon we described a few years ago as 'digital materiality'.[1] As a consequence, a uniform technological basis for architecture has been established, which from the onset of building industrialisation in the early 20th century was more vision than reality. We are no longer witnessing the delayed modernisation of an industry, but rather a historical departure: the modern division between intellectual work and manual production, between design and realisation, is being rendered obsolete.[2]

At the same time, a wide range of inherently architectural topics are finding their way back on to the agenda, not least among which are crafts and the art of construction, and, in particular, methods of architectural design. As robotics becomes increasingly commonplace in architecture, the subject of debate can no longer be its 'dematerialisation into pure form', as raised by digital architecture during the 1990s. Instead, what we are observing today is the comprehensive digitalisation of architecture, which entails a radical paradigm shift in its production conditions. The employment of robotics in architecture is opening up the prospect of entirely new aesthetic and functional potentials that could fundamentally alter architectural design and the building culture at large.[3]

Fabio Gramazio and Matthias Kohler, *Structural Oscillations*, Architecture and Digital Fabrication, ETH Zurich, Venice Architecture Biennale, 2008
On-site robotic fabrication significantly expands the range of digital manufacturing in architecture.

For the *Structural Oscillations* installation, a 100-metre (328-foot) long wall made of 26 segments with 14,961 overall individually rotated bricks was robotically fabricated.

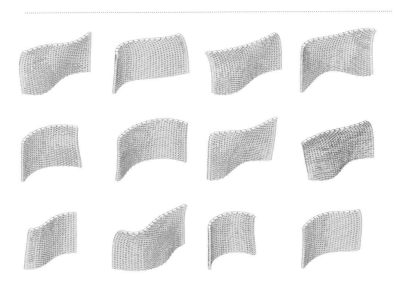

Robots

Employing digital technologies solely for the purpose of generating complex geometries and renderings warrants some scepticism, and there are plenty of cases in point. However, the question is whether digital technologies can impact and therefore change architectural material practice. Our research work with robots at ETH Zurich and the Future Cities Laboratory (FCL) at the Singapore ETH-Centre for Global Environmental Sustainability (SEC),[4] is heavily anchored in this voyage of discovery, and this issue of △ explores what happens if architecture absorbs the proposed connection – enabled by robots – between computational logic and material realisation as a new basis for the discipline's practices.

The industrial robot, because of its ability to perform an unlimited variety of non-repetitive tasks, is considered as the enabler for this deep transformation. However, rather than focusing on the technological development of robots itself, no matter how fascinating this might be, we are interested in establishing an architectural perspective by exploring the potential of robot-induced design and materialisation processes. To this end, we have reverted to using articulated-arm robots as established, cost-efficient fabrication machines that are at once both reliable and flexible, and whose potential in conventional industrial applications has been thoroughly proven. It is essential that architecture and the conditions specific to its production inform our approach to robotic fabrication, and not vice versa.[5] Only in this way can we significantly expand the range of architectural design and production options, enabling a new material differentiation and complexity to emerge and find expression.

Fabio Gramazio and Matthias Kohler, Complex Timber Structures, Architecture and Digital Fabrication, ETH Zurich, 2013
The use of a robot makes it possible to radically 'inform' material construction processes and to empower complex spatial assemblies from a large number of small elements.

The robot is able to automatically place and manipulate material according to a precise digital blueprint, and aggregate it under the explicit guidance of a computational architectural design.

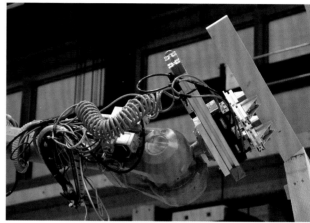

Start, Crash, Reboot

For various reasons, the initial introduction of robotic systems to the building construction industry was anything but a success. Most of all, the development of robot-based construction processes in the 1990s frequently led to either highly specialised, extremely expensive construction robots with extremely limited flexibility, or to robot-based construction factories yielding the same constraints.[6] Robots in building construction up to this point were used exclusively to further optimise (standardised) building processes, as a means of achieving only greater productivity (see the article on 'Changing Building Sites' by Thomas Bock and Silke Langenberg on pp 88–99 of this issue). Ultimately, no real lasting added architectural value, let alone any new (digital) building culture took hold during these initial attempts at integrating robotics into the building industry.

All of this radically changed at the turn of the millennium. Digital technologies became more commonplace among the architectural discipline and began having a greater impact on the understanding of architectural design and practice.[7] In addition, with the rapid spread of computer-controlled production machinery borrowed from other industries, such as milling and laser cutters or 3D printers, the 'digital project' attained considerable material value. In 2005, in order to examine the resulting new production conditions for architecture, a multipurpose fabrication laboratory employing an industrial robot – the first such laboratory in the field of architecture – was installed at ETH Zurich.

Industrial robots are distinguished by their versatility. Like computers, they are suitable for a wide variety of tasks because they are 'generic' and therefore not tailored to any particular application. Instead of being restricted in their operations to a prescribed range of applications, the 'manual dexterity' of robots can be freely designed and programmed. Their material manipulation skills can be customised to suit a specific constructive intention, both at the material and conceptual levels. It is precisely this quality – unleashing a previously unimaginable range of freedom in the interplay between the machine and the object – that distinguishes the operational applicability of industrial robots from all other specialised digital fabrication machines. In order to exploit this potential, which massively expands the concept of architectural design, not only a technical grasp of the robot's construction capabilities, but also an in-depth understanding of the materials to be processed, is necessary.[8]

Industrial robots are distinguished by their versatility. Like computers, they are suitable for a wide variety of tasks because they are 'generic' and therefore not tailored to any particular application.

Fabio Gramazio and Matthias Kohler, Robotic research facility, Architecture and Digital Fabrication, ETH Zurich, 2005
The world's first fabrication facility for employing industrial robots in architecture.

Challenging Scale

Only once digital architecture has assumed a more radical, substantial role in the aesthetic *and* material realisation of architecture will the discipline finally arrive in the digital age. The Gantenbein vineyard facade (Fläsch, Switzerland, 2006) represents the first of our projects to demonstrate this immanent architectural potential. It is remarkable because it anticipated the central principle of an additive robotic fabrication of architecture at full scale by demonstrating the non-standardised assembly of the extraordinarily large number of single (brick) elements.

We have since further intensified our explorations and sharpened our focus through a number of research projects. As such, the range of robotic processes is gradually expanding, from prefabrication towards direct use of robots on the construction site (see Volker Helm's 'In-Situ Fabrication' article on pp 100–107). *Flight Assembled Architecture* (FRAC Centre, Orléans, 2011) ultimately demonstrates, five years after the Gantenbein vineyard facade, that future robotically facilitated construction processes are capable of vastly surpassing previous scales of digital (pre-) fabrication.[9] In this architectural installation, several autonomously flying quadrocopters were employed to collaboratively assemble over 1,500 building modules to form a porous, vertical (urban) aggregation. In contrast to conventional building processes, here the flying robots were able to operate freely in airspace and amalgamate temporarily with the building materials they ultimately deposit.[10]

All of our projects are driven by a curiosity to explore the capabilities of robotic fabrication in relation to real building construction, with all its requirements and challenges. Rather

Gramazio & Kohler in cooperation with Bearth & Deplazes, Gantenbein vineyard facade, Fläsch, Switzerland, 2006
below top: The plasticity and hue of the rotated fields of bricks changes depending on the position of the sun, while from close up the three-dimensional depth dissolves and disappears in the detail of the individual bricks.

below bottom: The 400-square-metre (4,300-square-foot) facade of the Gantenbein Winery was robotically fabricated from 20,000 individually rotated bricks.

than a mere theoretical exercise, we regard this empirical attitude as crucial to unlocking the full potential of robotic practices in architecture. This materialist approach proceeds from an understanding of design that is directly informed by the material's constructive capacities in conjunction with well-attuned fabrication principles (see Norman Hack and Willi Viktor Lauer's research on Mesh-Mould structures on pp 44–53). In turn, rather than merely 'illustrate' a predetermined design idea, architectural design should be informed by novel fabrication processes directly derived from the logic of the given material. As architecture embodies pure physical substance, the concrete task is to figure out creatively – together with robots – the relevant material processes at hand. Here, robotic fabrication brings novel design concepts and performative capacities of materials to the foreground (see Neri Oxman's investigation into 'robotic swarm printing' on pp. 108–15).

This issue of △ looks at new digital fabrication paradigms and provides a basis for contextualising and examining the scope of this focal point in detail. It presents selected contributions that not only challenge the reputedly clear division between theory, research and practice, but also serve as an instrument for sounding out the application of robotics at full architectural scale. As the title of the issue suggests, the driving force behind these explorations is the desire to seek lasting future strategies for 'making' innovative architecture with robots.

Venturing Out of Bounds

Our core contribution to the issue, the Design of Robotic Fabricated High Rises design research studio at the Future Cities Laboratory illustrates a pioneering attempt to place digital fabrication in the context of architectural production, and to explore the potential of robotic construction processes in the context of large-scale residential tower developments. In order to overcome the prevailing paradigm of repetition and mono-functionality in such urban developments, as well as the resulting monotony, the central concern of the project is the tectonic inquiry into high-rise typologies through digital design and fabrication processes. The design research studio in Singapore is geared towards 1:50 models of mixed-use high-rises, which are computationally designed and robotically fabricated (see Michael Budig, Jason Lim and Raffael Petrovic's contribution 'Integrating Robotic Fabrication in the Design Process' on pp 22–43).

Hereby, robotic fabrication overcomes the repetitive build-up of standard building elements in favour of a differentiated assembly of bespoke elements, and links computational design to the fabrication of physical study models. Our thesis is that through such hybrid digital-physical methods, the physical model – even in the age of computation – again gains central significance.[11]

Robotics Pioneers

Is the robot also a new El Dorado for startups? In addition to academic research and teaching projects, *Made by Robots* presents important pioneers in this field: a selection of startups devoted entirely to architectural robotic fabrication processes (see pp 60–75). They are distinguished by their nonchalance as well as their savvy and innovative application of novel production technologies. At Odico Formwork Robotics, RoboFold, Machineous or ROB Technologies and GREYSHED, the usual coveted symbols of status, brand identity and recognition are being trumped by the advantage

Gramazio & Kohler and Raffaello D'Andrea in cooperation with ETH Zurich, *Flight Assembled Architecture*, FRAC Centre, Orléans, France, 2011

Leaving the common workspace of conventional digitally controlled production machines, flying robots can operate freely in airspace. Here they assemble collaboratively over 1,500 building modules into a vertical urban structure.

Even if today's very specific implementation of robotic fabrication processes – in comparison to the lack of material substance in the early days of architecture's digitalisation – is practically forcing architecture's arrival in the digital age, what still remain vague are the ties to architectural history, its theoretical implications and future prospects.

in being curious, adaptable and versatile. An entrepreneurial spirit that favours cooperation over authorship pervades these enterprises, in the sense of empowering oneself and others.[12] Each is striving to remove the stigma associated with the mechanical-age mentality of standardised thinking and serial production. Here, robots offer a reliable and cost-effective technology that is globally accessible and extremely flexible in its application (see particularly the text by Machineous on pp 70–71).

It is this understanding that underlies the focus of the startups featured in this issue, not only on the programming and application of robotic fabrication processes – the assembly of non-standardised building elements (ROB Technologies, pp 72–3), the cutting of individual formwork (Odico Formwork Robotics, pp 66–7) or the folding of sheet metal (RoboFold, pp 68–9) – but also on the development of novel interfaces, software or design processes. Academic research in this area is giving rise to new enterprises and innovative business ideas, and comparable to the 3D printing sector their dynamism is increasingly taking hold and permeating the entire field of architectural activity to the hum of 'how to make almost anything'.[13] Should robotic fabrication processes actually become commonplace in the construction industry over the next decades, these 'pioneers' could be credited with having dauntlessly transformed the building industry bottom-up (see GREYSHED, pp 74–5 and facilitated the breakthrough of the digital architectural production of the future.

Theory of Change – Change of Theory

Even if today's very specific implementation of robotic fabrication processes – in comparison to the lack of material substance in the early days of architecture's digitalisation – is practically forcing architecture's arrival in the digital age, what still remain vague are the ties to architectural history, its theoretical implications and future prospects. Rather than run the risk of getting caught up in the latest robotic technologies, a reflexive awareness is called for that asks: How do we define architectural production in today's world, if the current dichotomies of code and material, type and variation, author and artefact, human and machine, are increasingly losing ground and being replaced by new categories that render them outdated? (See, for example, the work of François Roche/New-Territories on pp 116–25.)

What is the point of architectural discourse today if the dominant narrative finds itself increasingly clutching at straws rather than contributing to greater clarity? Neither the pledge to overcome the Modernist legacy nor the sole focus on a (post-) digital future is of much use in this case.[14] Explorations into our current digital age must broaden to include both practical *and* theoretical perspectives. This would require that the robot be regarded not only as a medium of production, but also as an epistemological approach (see Antoine Picon's encounter with the theoretical implications of architectural robotics on pp 54–9). Only then would it be possible to stimulate relevant and rigorous exploration, discussion, principles and prospective applications for robotics in architecture.

Architecture in the Second Digital Age

This issue of *D* ventures to take a look forward because we believe robotic potential in architecture is inextricably future-oriented. The robotic fabrication of tomorrow, surprisingly,

Fabio Gramazio and Matthias Kohler, Design of Robotic Fabricated High Rises, Architecture and Digital Fabrication, Future Cities Laboratory (FCL), Singapore-ETH Centre for Global Environmental Sustainability (SEC), 2013

opposite: The robotic fabrication of digitally generated 1:50 models of mixed-use high-rises in an integral design research studio places digital manufacturing in an urban context and allows the experimental connection of algorithmic design to the physical act of building.

below: As the Design of Robotic Fabricated High Rises project shows, the use of robots in architecture allows the connection of imagination and construction like never before, in its potential to reveal a radically new way of thinking about and materialising architecture.

will no longer be bound to constricting standards, constraints or ideologies, but will allow each architectural experiment with robots the freedom to follow its own agenda. We now have access to enormous knowhow and different forms of knowledge – anyone can become an expert in digital fabrication these days. The present moment is thus ripe for revolutionising architectural production; robots are now connecting technology and knowhow, as well as imagination and materialisation, like never before, and have the potential to reveal a radically new way of thinking about and materialising architecture. This takes away the abstract and forced artificial character of the digital in architecture and imbues it with a totally distinct material significance and identity. One could even speak of the dawn of a 'second digital age'. This issue hereby signals a seminal shift. Or, in other words, architecture is at long last beginning to develop an adequate material practice for the cultural logic of the information age.

Acknowledgement

We would like to express our gratitude to Helen Castle for entrusting us with an △ edition. We extend our thanks as well to all of the authors who contributed to making this issue of △ exceptional. Above all, we would like to thank Silke Langenberg for her invaluable support and unparalleled commitment, as well as Caroline Ellerby and Orkun Kasap for helping us prepare this issue. Our sincere appreciation goes out as well to our collaborators, students and project partners, both in Zurich and in Singapore, who have significantly enriched and inspired our work over the course of many years. △

Notes
1. Fabio Gramazio and Matthias Kohler, *Digital Materiality in Architecture*, Lars Müller Publishers (Baden), 2008, pp 7–11.
2. Mario Carpo, 'Revolutions: Some New Technologies in Search of an Author', *Log 15*, Anyone Corporation (New York), 2009, pp 49–54.
3. Fabio Gramazio, Matthias Kohler and Jan Willmann, *The Robotic Touch: How Robots Change Architecture*, Park Books (Zurich), 2014.
4. The Future Cities Laboratory (FCL) is a transdisciplinary research centre for urban design and sustainability at a global scale, under the direction of the Singapore-ETH Centre for Global Environmental Sustainability (SEC). See: www.futurecities.ethz.ch.
5. Tobias Bonwetsch, Fabio Gramazio and Matthias Kohler, 'Towards a Bespoke Building Process', in Bob Sheil (ed), *AD Reader: Manufacturing the Bespoke – Making and Prototyping Architecture*, John Wiley & Sons (Chichester), 2012, pp 78–87.
6. Undoubtedly, a prime example is the custom bricklaying machine ROCCO (Robot Construction System for Computer Integrated Construction). See Thomas Bock and Friedrich Gebhart, 'ROCCO – Robot Assembly System for Computer Integrated Construction: An Overview', *First European Conference on Product and Process Modelling in the Building Industry*, Dresden, 1994, pp 5–7.
7. Bart Lootsma, 'Foreword', in Peter Zellner (ed), *Hybrid Space/Digital Architecture*, Thames & Hudson (London), 1999, pp 7–16.
8. Jan Willmann, Fabio Gramazio, Matthias Kohler and Silke Langenberg, 'Digital by Material: Envisioning an Extended Performative Materiality in the Digital Age of Architecture', in Sigrid Brell-Cokcan and Johannes Braumann (eds), *Rob|Arch 2012: Robotic Fabrication in Architecture, Art and Design*, Springer (Vienna), 2012, pp 12–27.
9. Jan Willmann, Fabio Gramazio and Matthias Kohler, 'The Vertical Village', in Fabio Gramazio, Matthias Kohler and Raffaello D'Andrea (eds), *Flight Assembled Architecture*, Editions HYX (Orléans), 2013, pp 13–23.
10. Matthias Kohler, 'Aerial Architecture', *Log 25*, Anyone Corporation (New York), 2012, pp 23–30.
11. Jason Lim, Fabio Gramazio and Matthias Kohler, 'A Software Environment For Designing Through Robotic Fabrication', *Open Systems: Proceedings of the 18th International Conference on Computer-aided Architectural Design Research in Asia (CAADRIA)*, Hong Kong, 2013, pp 45–54.
12. Mario Carpo, 'Digital Darwinism: Mass Collaboration, Form-Finding, and the Dissolution of Authorship', *Log 26*, Anyone Corporation (New York), 2012, pp 97–105.
13. See Neil Gershenfeld, *The Coming Revolution on Your Desktop: From Personal Computers to Personal Fabrication*, Basic Books (New York), 2007.
14. Antoine Picon, 'The Ghost of Architecture: The Project and Its Codification', *Perspecta 35: Building Codes*, MIT Press (Cambridge, MA), 2004, pp 9–19.

Text © 2014 John Wiley & Sons Ltd. Images: pp 14–16, 17(t), 20, 21 © Gramazio & Kohler, Architecture and Digital Fabrication, ETH Zurich; pp 17(c&b) © Ralph Feiner; p 18 © FRAC Centre, photography François Lauginie

Michael Budig, Jason Lim and Raffael Petrovic

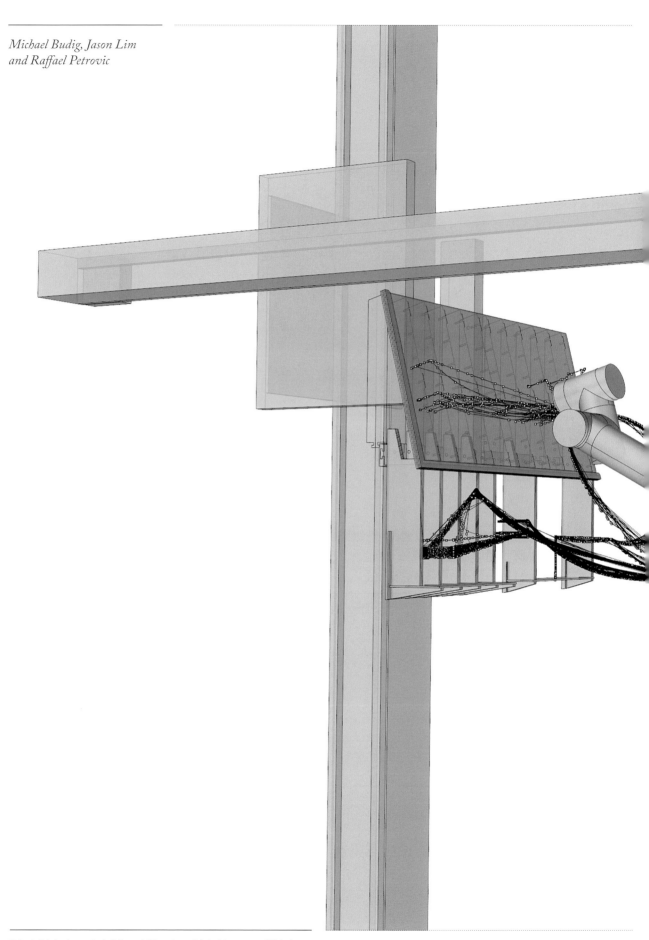

Robotic fabrication paths building a 1:50-scale model, Architecture and Digital Fabrication, Future Cities Laboratory (FCL), Singapore-ETH Centre for Global Environmental Sustainability (SEC), 2012

INTEGRATING ROBOTIC FABRICATION IN THE DESIGN PROCESS

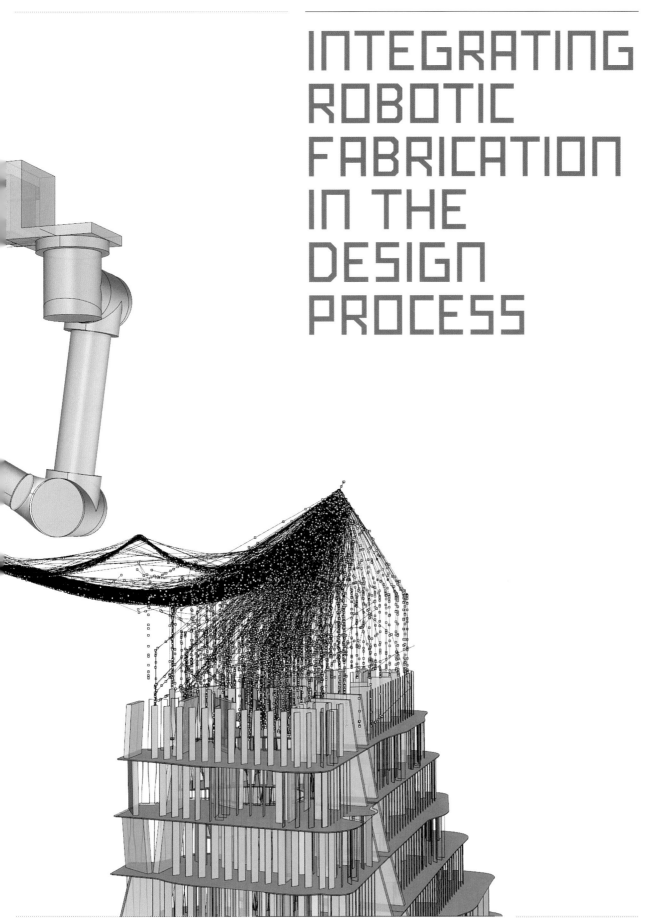

The production paths can be anticipated to avoid possible collisions of the robotic arm with the already assembled components, and actual paths can be recorded to check for possible optimisation in the building process.

Does the application of robotic technologies in Southeast Asian cities, where the residential high-rise is pre-eminent, provide a unique opportunity to introduce bespoke design elements and liberate the high-rise from serial production and standardisation? **Michael Budig, Jason Lim and Raffael Petrovic** of the Future Cities Laboratory (FCL) at the Singapore-ETH Centre for Global Environmental Sustainability (SEC) introduce the research project 'Design of Robotic Fabricated High Rises' (2012–13) led by Fabio Gramazio and Matthias Kohler, exploring high-rise typologies and architectural scenarios for future vertical cities.

Fast-growing regions in Southeast Asia and China face a continual demand for housing. In cities like Singapore, high-rises represent the most common residential typology. Around 80 per cent of Singapore's population live in flats, which have been erected by the Housing Development Board (HDB). Similarly to Hong Kong, these programmes were initiated from the 1950s onwards to provide housing for millions of people in a very short time. High-rises foster a strictly repetitive distribution of identical building elements along the vertical axis, and thus ideally fulfil the efficiency criteria of a still dominant industrial building paradigm. The result is very often reflected in overly uniform urban environments. The research project Design of Robotic Fabricated High Rises investigates the potentials of robotic building processes for the construction of this typology. The hypothesis is that robotic technologies in combination with computational design techniques liberate this widespread building typology from the limitations of a serial production paradigm. Professors Fabio Gramazio and Matthias Kohler lead the research project, which is one of 13 different modules of the Future Cities Laboratory (FCL) at the Singapore-ETH Centre for Global Environmental Sustainability (SEC).[1]

Robotic facilities, Architecture and Digital Fabrication, Future Cities Laboratory (FCL), Singapore-ETH Centre for Global Environmental Sustainability (SEC), 2012

The robotic facilitiies at the Future Cities Laboratory are used by masters-level and PhD students for research on digital fabrication processes at different scales.

The research is conducted on two distinct levels, which constantly inform each other. PhD researchers approach the subject from a scientific perspective, focusing on the integrated development of computation, fabrication and material systems. In parallel, a design research studio explores the impact of such changed production conditions on architectural design, with small groups of master's-level students designing parametric high-rise typologies and creating architectural scenarios for future vertical cities.[2] An experimental methodology is applied in the studio, whereby designs are explored using computational techniques and materialised through robotic fabrication. Two ideas are central. First, 1:50-scale models serve as the primary medium for design exploration. Second, rather than designing forms, the focus is set on designing processes, which are algorithmically described and robotically executed. The objective is to expand present computational design methodologies, which are already well established in academia, by introducing a material counterpart in the form of robotically fabricated physical working models.

The Role of the Physical Model in Computational Design

Computer-aided design (CAD) technology has advanced rapidly in the past two decades and is now widely adopted by architects. Today, digital models can be produced faster and cheaper than physical ones. Despite being represented on a flat screen, the digital model appears to be three-dimensional since it can be rotated, moved and navigated in real time. Furthermore, it can be rendered to produce photo-realistic still images. With these qualities, the digital model has undermined the primacy of the physical model as a representational device. At the same time, it has also largely replaced the working model, as it is easy to edit and thus even more suited for quick iterative design exploration than its physical counterpart.

While a digital model can potentially contain all construction details down to the smallest component, a scaled physical model can only embody a limited amount of information. However, despite this intrinsic limitation, physical models retain distinct advantages over digital ones. Working models provide architects with a direct and physical means to study and understand the three-dimensionality of their design, which cannot be fully grasped through a two-dimensional digital representation. In order to elaborate specific design ideas in a working model, architects have to choose an appropriate level of abstraction. While building them, they sharpen the key concepts of the design. In addition, the physical model immediately communicates the relationships between material and structure, space, and proportions. It provides architects with direct sensual and haptic feedback that is completely missing in a digital environment.

Thus, the thesis of the design research studio is to reposition the physical model as a critical explorative tool in conjunction with computational design, whereby robotic technology is used for its fabrication. With its ability to execute individualised actions and to position elements in space without external reference systems, the robot enables students to build models using complex assembly logics, out of a large number of parts and within a reasonable amount of time. In contrast to 3D-printed models, which are based on an undifferentiated, layer-based fabrication process, robotically fabricated models bring into focus structure and tectonic characteristics. In this regard, they offer to the designer the same kind of insights as their handmade equivalents. However, robotic fabrication creates a direct and rigorous link between the physical model and its computational origin.

A 1:50 model scale was chosen as it exposes explicit structural and constructive problems, and allows students to explore articulated architectural solutions. Standing up to 3 metres (10 feet) tall, the model towers in the design studio become 'buildings, with their own complex engineering and construction problems'.[3] They are evocative physical expressions of a design approach that combines abstract digital thinking with concrete tectonic sensibility.

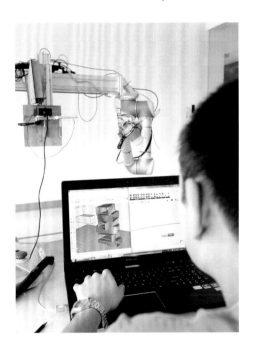

The thesis of the design research studio is to reposition the physical model as a critical explorative tool in conjunction with computational design, whereby robotic technology is used for its fabrication.

Student controlling the robotic fabrication process, Architecture and Digital Fabrication, Future Cities Laboratory (FCL), Singapore-ETH Centre for Global Environmental Sustainability (SEC), 2013

The robotic controlling components are embedded in a visual programming environment. The student checks the conceived model with the physically fabricated element and incorporates adjustments into the digital model.

The Design of the Robotic Process

The industrial robot has been designed as a general-purpose machine. Hence, it must be customised for a specific application, in this case model fabrication, through physical tooling of specific end effectors and peripheral tools. With a suitable end effector attached to its tip, a robot is able to manipulate material for a bespoke constructive process. The robotic facilities at the Future Cities Laboratory have been conceived as an open environment in which the robots can be extended by the students with a kit of modular parts, including grippers, material feeders and feedback sensors. This allows a wide range of design experiments to be supported. Unlike their larger industrial counterparts, the robotic arms installed in the design studio are safe to work with, enabling direct human–robot interaction. They are mounted on vertical axes, enlarging their working envelope and allowing for the fabrication of tall study models.

In addition to the physical tooling of end effectors, robots, like any computer-controlled machine, can be programmed in their movements. But the production of control data in the form of coded machine instructions requires programming skills as well as an understanding of robotics concepts such as kinematics, which lie outside most architects' domain of expertise. To address this problem, the custom robot programming library *YOUR*, which aims at making robot control intuitively accessible to students, has been implemented. Based on this, a toolkit of visual programming components for Grasshopper™, allowing students to directly assemble graphical components for the control of their robotic fabrication sequences, was developed. Due to its immediacy, this visual programming approach is well suited for quick prototyping of initial, simple robotic processes. As the text-based code defining these components is transparent and accessible, once students acquire experience in programming and knowledge in robotics, they become able to modify them in order to introduce more complex flow structures and logics into their robotic processes.

> Unlike their larger industrial counterparts, the robotic arms installed in the design studio are safe to work with, enabling direct human–robot interaction.

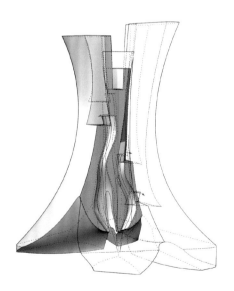 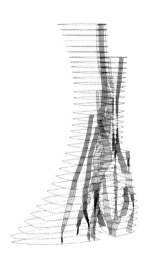 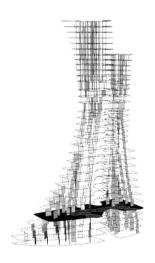

1. Generation of outer volume and inner voids
2. Creation of primary wall system
3. Generation of secondary wall system

Pascal Genhart and Tobias Wullschleger, Design of Robotic Fabricated High Rises, Architecture and Digital Fabrication, Future Cities Laboratory (FCL), Singapore-ETH Centre for Global Environmental Sustainability (SEC), 2012

this page and opposite bottom (both): Diagrams showing various layers of information in the digital model: void systems with public zones and circulation, structural system, and building components for the fabrication process.

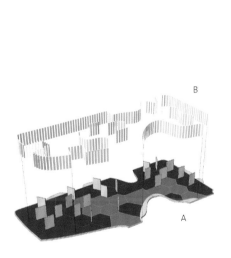

4. Population of interior walls with informed louvres

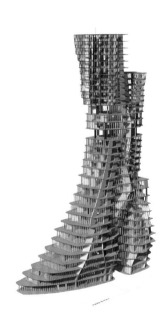

5. Final tower

Design of Robotic Fabricated High Rises, Architecture and Digital Fabrication, Future Cities Laboratory (FCL), Singapore-ETH Centre for Global Environmental Sustainability (SEC), 2012–13

A multitude of different models robotically fabricated in the design research studio in Singapore.

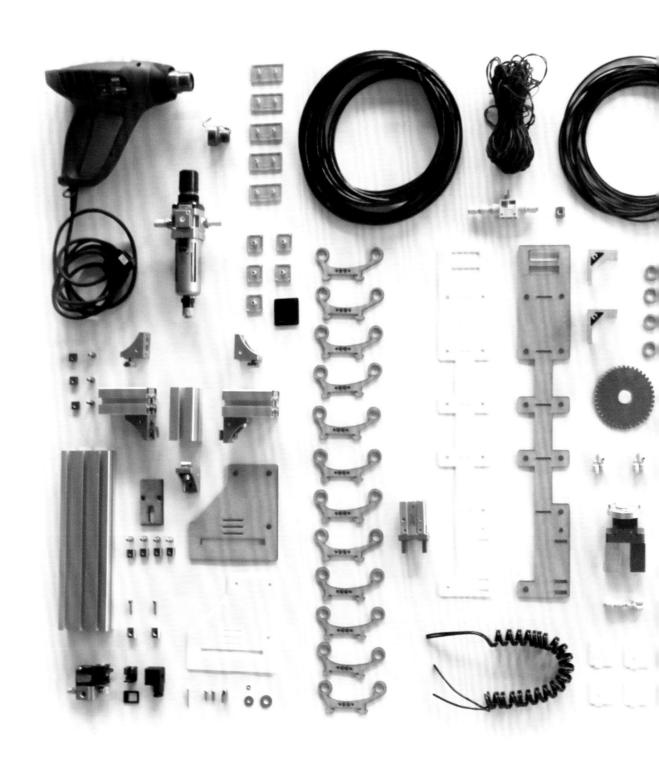

Sylvius Kramer and Michael Stünzi, Design of Robotic Fabricated High Rises, Architecture and Digital Fabrication, Future Cities Laboratory (FCL), Singapore-ETH Centre for Global Environmental Sustainability (SEC), 2012

Dismantled setup of a robotic end effector, developed for bending acrylic strips through thermal deformation. Students were able to interact directly with the robotic systems, adjust the end effectors to their needs, and eventually invent setups for various fabrication and assembly processes.

Robotic fabricated models at the robotic facilities after completion, Architecture and Digital Fabrication, Future Cities Laboratory (FCL), Singapore-ETH Centre for Global Environmental Sustainability (SEC), 2013

Each student team had one robotic facility for the development of their designs, and for the fabrication of up to five generations of 1:50-scale models.

  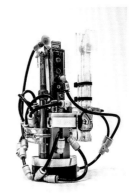 

Sebastian Ernst, Sven Rickhoff and Silvan Strohbach, Design of Robotic Fabricated High Rises, Architecture and Digital Fabrication, Future Cities Laboratory (FCL), Singapore-ETH Centre for Global Environmental Sustainability (SEC), 2012

Series of robotic end effectors were developed for various materials and constructive systems. The illustrated gripper was conceived for a robotic stapling process, which clamps two paper strips. Since the two paper strips can be pushed forward with varying speed between two clamping locations, the material is forced to bend, producing curvilinear shapes.

The Integration of Computation and Fabrication

The design research studio sets the emphasis on designing processes, which integrate design computation and robotic fabrication, as opposed to directly designing forms. Students develop their architectural concept along with a specific computational design strategy and a custom robotic fabrication process. While the parallel development of computation strategies and fabrication processes sharpens the students' awareness for their conceptual interdependence, their synthesis results in digitally driven yet tectonically informed designs.

In order to become able to computationally formalise their ideas by means of either visual programming or scripting, students first have to identify the relational logics underlying their architectural concepts. The resulting computational models do not only encapsulate the design logic, but also incorporate the procedure for their digital robotic fabrication. Thereby, the students are not only able to generate and visualise the geometry of their design, but can also produce the control data needed by the robot for their (scaled) digital production.

Robotic fabrication and computational design augment each other in several ways. In addition to the direct materialisation of digitally conceived designs, the physical working model provides students with a differentiated and sensual feedback on their design, whereby they can immediately engage with its constructive and structural aspects. As part of an iterative process, these insights can then be re-integrated into the computational design strategy.

Once the connection between the computational model and the robotic fabrication process is operational, this design methodology allows for the stepwise refinement of the design through the sequential fabrication of multiple working models. By iteratively refining the rules and tuning the parameters of the computational model throughout the design development, multiple versions can be physically evaluated and compared.

As the constructive capabilities of the robotic fabrication process clearly define the design space and thus productively inform the computational design strategy, its development can also be recognised as a creative act of design on its own.

Potentials of Robotic Technologies for Design Practice

The discussed methodology was tested over two consecutive one-year-long master's-level design research studios. The results, some of which are featured over the following pages of this issue, demonstrate the potential of the integration of robotic fabrication technologies into the computational design process of large-scale architectural typologies. This methodology directly addresses the shortcomings of contemporary digital design practices that overly privilege computation, by offering an alternative that stresses its complementary relation to physical (model) making.

The achievements of the studio were wide-ranging: compelling architectural concepts were developed, complex physical models were built and bespoke fabrication processes were engineered. Though less tangible, the most significant result is to be found in the increasing awareness of computational logics among the students exposed to this design methodology. After acquiring the skills to productively use computation and digital fabrication for design purposes, by developing their projects they learned to synthesise their thoughts in algorithmic logics and to translate them into material, constructive processes.

Architectural design practice will be increasingly mediated by digital technology in the future. Digital fabrication technology allows architects to conceive designs both digitally and physically, and may empower them to take a more active role in the materialisation and construction process.

As a consequence, students today must become acquainted with such technologies as part of their education. Only then will these architects-in-the-making be equipped with the technical and, more crucially, intellectual skills, to navigate and actively influence this new technological landscape.

Notes
1. The SEC is co-funded by the Singapore National Research Foundation (NRF) and the Swiss Federal Institute of Technology Zurich (ETH Zurich) Department of Architecture.
2. The design studio was conducted twice in 2012 and 2013 as a two-semester programme with students from both ETH Zurich and the National University of Singapore (NUS). It was conducted by Matthias Kohler and Fabio Gramazio with Willi Viktor Lauer, Norman Hack and the authors.
3. Jane Jacobs, 'The Miniature Boom', *The Architectural Forum*, May 1958, p 107.

As the constructive capabilities of the robotic fabrication process clearly define the design space and thus productively inform the computational design strategy, its development can also be recognised as a creative act of design on its own.

NESTED VOIDS

Pascal Genhart and Tobias Wullschleger, Nested Voids, Design of Robotic Fabricated High Rises, Architecture and Digital Fabrication, Future Cities Laboratory (FCL), Singapore-ETH Centre for Global Environmental Sustainability (SEC), 2012

The final 1:50-scale model is 2.7 metres (9 feet 10 inches) high and consists of several thousand uniquely cut and placed cardboard elements, and heat-deformed acrylic louvres.

The basic architectural idea of this project was to organise a high-rise through a system of interconnected voids, which contain the tower's circulation zones and shared spaces, provide natural ventilation and create visual connections between the floors. The design comprises a cluster of interacting towers, incorporating a railway station on its ground levels.

The integration of non-standard prefabrication and robotic assembly in an enhanced workflow enlarged the design opportunities, as the production and assembly of wall and floor elements have no longer been based on repetition. To control the resulting differentiated construction process, all the data for design, laser cutting, material deformation, and robotic assembly were generated in the computational environment. Later the robotic setup has been extended with a hot-air gun for thermal deformation, which was used to twist acrylic strips for a sophisticated louvre facade system.

The section shows the tower's branching system of concrete walls, which is generated by an algorithm correlating to the interconnecting void spaces.

Pascal Genhart, Patrick Goldener, Florence Thonney and Tobias Wullschleger, Nested Voids, Design of Robotic Fabricated High Rises, Architecture and Digital Fabrication, Future Cities Laboratory (FCL), Singapore-ETH Centre for Global Environmental Sustainability (SEC), 2012

The robotic fabrication process is based on the integration of non-standard element prefabrication and robotic assembly. The model's walls and floors are laser cut out of cardboard sheets, which are placed on a custom-designed feeder system (left). From there the robot picks the elements (centre), applies glue and places them directly on the model (right).

SEQUENTIAL FRAMES

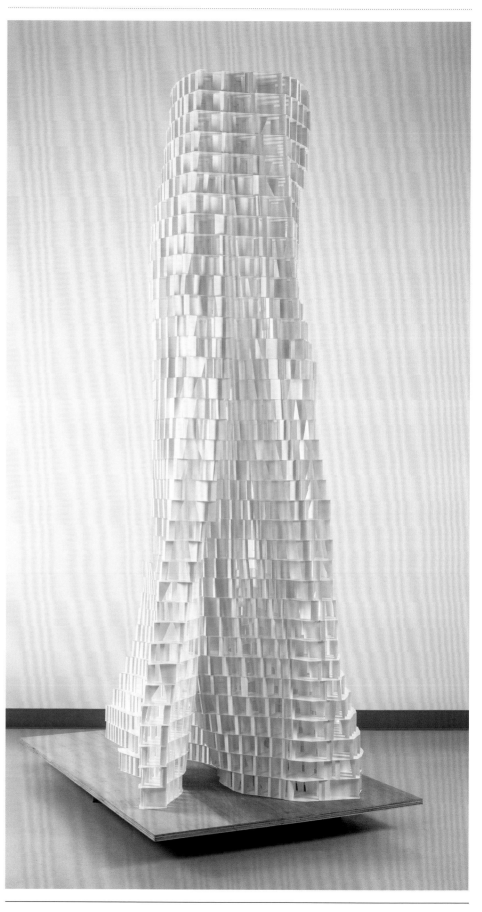

David Jenny and Jean-Marc Stadelmann, Sequential Frames, Design of Robotic Fabricated High Rises, Architecture and Digital Fabrication, Future Cities Laboratory (FCL), Singapore-ETH Centre for Global Environmental Sustainability (SEC), 2013

The tower, 3 metres (10 feet) in height, is assembled out of 5,200 robotically cut and folded paper elements.

The design intention of this tower is to create a multitude of unique interior spatial experiences out of simple geometrical elements, by deploying the full power of computation. The towers are planned as linked strands that branch and merge into an undulating overall shape, bridging a Singaporean highway and connecting two adjacent parks.

Usually high-rises are designed as envelopes and then sliced into floors. This traditional approach is inverted by designing a tower starting from its interior spaces. In this project the number of shear walls was increased, and simple rules were experimented with to define their internal organisation. The design code distributes predefined openings into sequences of the shear walls to accommodate different flat types. In a second step the continuous force flow around these apertures is calculated from top to bottom. Each wall adjusts its opening's geometry, negotiating between the required structural performance and the desired cut-out for the flats, to make each flat appear different, even if they are all of the same type.

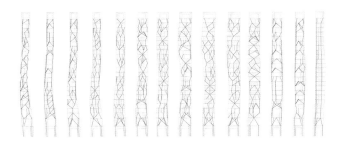

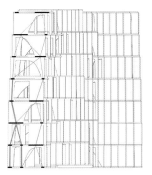

top and centre: Walls no longer frame a room but only a function. Interior spaces are not defined by enclosure, but by programmed cut-outs in a sequence of densely placed walls. The cut-outs follow the structural force flow.

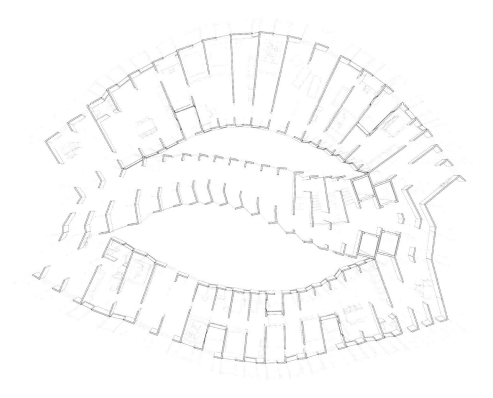

bottom: By defining simple algorithmic rules for the distribution of openings in the shear walls, the design engine creates a high-rise with a wide variety of unique interior spaces.

VERTICAL AVENUE

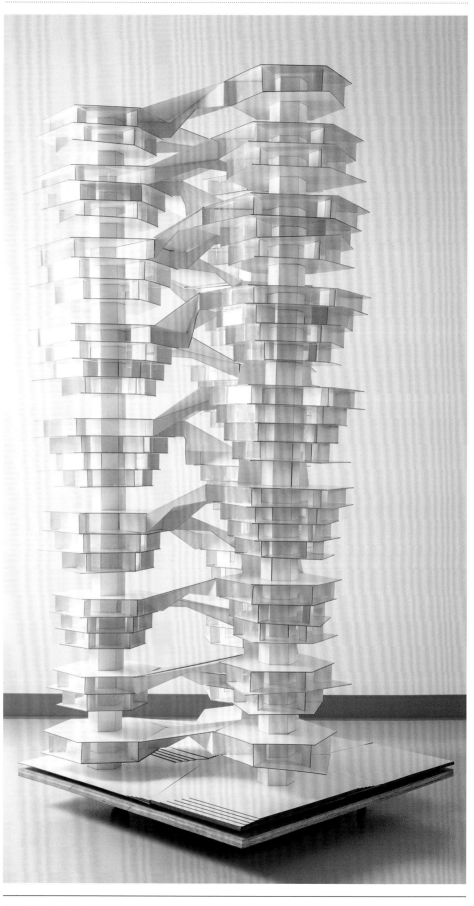

Foong Kai Qi and Kan Lijing, Vertical Avenue, Design of Robotic Fabricated High Rises, Architecture and Digital Fabrication, Future Cities Laboratory (FCL), Singapore-ETH Centre for Global Environmental Sustainability (SEC), 2013

The building consists of three high-rise towers with apartment clusters around the central cores. A system of staggered open public spaces is connected by a continuously upward-spiralling ramp, which serves as a public meeting space and park.

This project proposes a spiralling circulation system that provides public programmes and parks vertically throughout the tower. The idea is based on so-called 'void decks' – ground floors of Singaporean high-rise housing that are usually left open and used as public meeting spaces. The apartments are organised as clusters around three high-rise towers with vertically staggered void decks, that are connected by a continuously upward-spiralling ramp.

Apartments and facade elements are treated differently according to their positions in the towers. Flats oriented to the outside have cantilevering slabs that provide terraces and act as a shading system. All facades are set back to protect the units from Singapore's equatorial sun. Flats on the inside are enclosed with curved facade panels in order to deflect views and ensure certain degrees of privacy. All apartments have direct access to the nearest void deck.

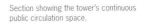

Section showing the tower's continuous public circulation space.

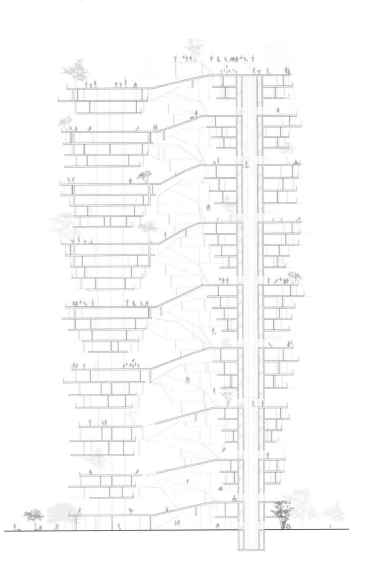

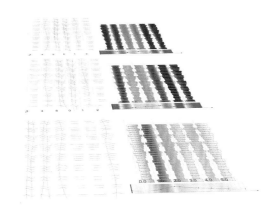

The tower was assembled using sensor-assisted picking and placing, in combination with robotically controlled thermal bending of acrylic facade elements.

MESH TOWERS

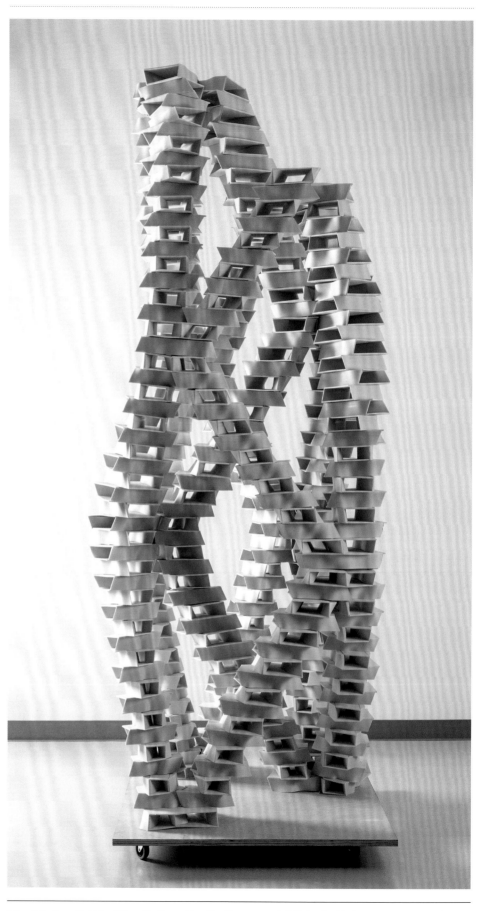

Petrus Aejmelaeus-Lindström, Chiang Pun Hon and Lee Ping Fuan, Mesh Towers, Design of Robotic Fabricated High Rises, Architecture and Digital Fabrication, Future Cities Laboratory (FCL), Singapore-ETH Centre for Global Environmental Sustainability (SEC), 2013

Individual residential units are generated on top of each other and oriented away from their immediate neighbours to provide privacy. The thickness of the load-bearing walls is locally adapted to the vertical force flow.

This project proposes a porous mesh of slender tower strands as an alternative to a massive condominium project, planned in direct neighbourhood to a conservation area of Singapore. The footprints of the overall shape are minimised, keeping the ground level open for a park and complementing the historical structure of its surroundings. The towers' nodes contain public programmes, such as restaurants, event spaces and kindergartens. Thin bridges connect these shared spaces, creating a hiking trail in the city.

For the final tower model, foam blocks are placed on a specially designed feeder. The robot cuts each of them differently, optimised to transfer the loads and shaped to accommodate various functions. This process in particular takes advantage of the robot's capability to accurately move and orient in space.

The robot picks the piece and moves it through a hot wire along a computationally generated path to fabricate each wall element.

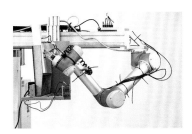

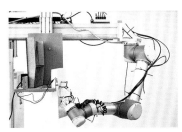

The footprints of the overall shape are minimised, keeping the ground level open for a park and complementing the historical structure of its surroundings.

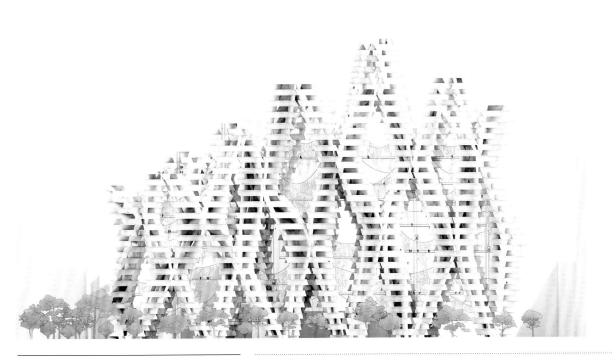

The project consists of several slender towers merging and separating as they grow in height, structurally supporting each other.

BENT STRATIFI- CATIONS

Sylvius Kramer and Michael Stünzi, Bent Stratifications, Design of Robotic Fabricated High Rises, Architecture and Digital Fabrication, Future Cities Laboratory (FCL), Singapore-ETH Centre for Global Environmental Sustainability (SEC), 2012

The bent acrylic strips are stacked on each other, in order to create the tower's primary structure.

This high-rise building is designed as a cluster of three mixed-use towers, stabilised by leaning against each other. The project creates a new hub for an urban development area in the east of Singapore. It proposes a three-dimensional web of circulation and public spaces between the towers. The concept builds upon horizontal stratification and provides differentiated apartment types, including split-level.

The processes in this project were based on working with a high number of identical, small building components. The amount of elements thus shifted the focus to investigating more complex aggregation logics. The final model still consists of generic basic elements, but the standard acrylic strips are thermally deformed into unique geometries during the robotic assembly process. This simplified the preproduction of the elements, and enhanced the picking and placing process while robotically informing their shapes.

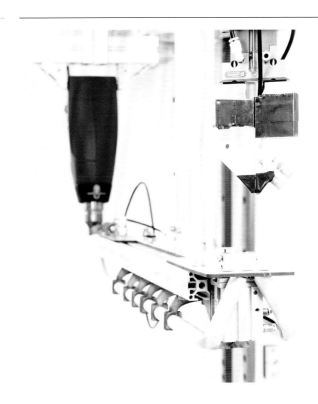

The processes in this project were based on working with a high number of identical, small building components.

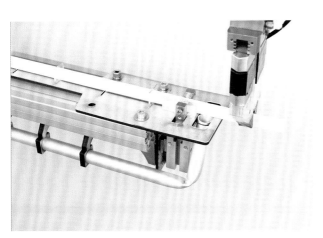
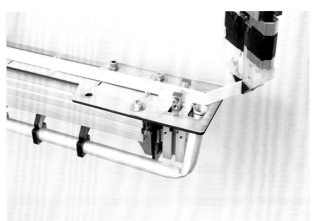

top and bottom: Acrylic strips are fixed in a linear rail and pulled forward to a designated bending position. The material is heated up and then bent by the robot to any desired angle.

UNDULATING TERRACES

Sebastian Ernst, Sven Rickhoff and Silvan Strohbach, Undulating Terraces, Design of Robotic Fabricated High Rises, Architecture and Digital Fabrication, Future Cities Laboratory (FCL), Singapore-ETH Centre for Global Environmental Sustainability (SEC), 2012

The final tower investigates a robotic stapling process of paper strips to create the primary structure and form of the tower.

Residential units are organised between a large internal void and an exterior undulating facade containing terraces. The walls of the inner void act as the tower's primary structural core, supported by laminating and delaminating layers of walls as secondary structure.

After numerous quite different concepts, the final model was built using the robot to staple together paper strips at various positions. The resulting curved paper elements were used to produce the facades and the exterior corridors of the tower. The robot's ability to precisely staple connections in varying distances formed a synergy with the inherent material properties to enable complex geometric shapes. This project illustrates the robot's versatility to produce a wide variety of designs, and how to utilise the feedback loop between the physical model and the digital design. ᴆ

Sebastian Ernst, Sven Rickhoff, Silvan Strohbach and Martin Tessarz, Prestudies to Undulating Terraces, Design of Robotic Fabricated High Rises, Architecture and Digital Fabrication, Future Cities Laboratory (FCL), Singapore-ETH Centre for Global Environmental Sustainability (SEC), 2012

The first iterations of towers were fabricated with simple picking and placing processes of modular components. In a second generation, material deformation was introduced, with custom-designed cardboard and aluminium bending processes.

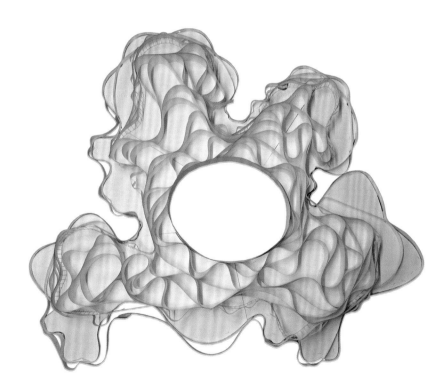

Sebastian Ernst, Sven Rickhoff and Silvan Strohbach, Undulating Terraces, Design of Robotic Fabricated High Rises, Architecture and Digital Fabrication, Future Cities Laboratory (FCL), Singapore-ETH Centre for Global Environmental Sustainability (SEC), 2012
One floor layer of the high-rise model showing the robotic stapled paper strips.

Text © 2014 John Wiley & Sons Ltd. Images: pp 22-3, 25, 26-7, 28-9, 30(b), 32-43 © Gramazio & Kohler, Architecture and Digital Fabrication, ETH Zurich; p 24 © Gramazio & Kohler, Architecture and Digital Fabrication, ETH Zurich, photo Bas Princen; p 30(t) © Gramazio & Kohler, Architecture and Digital Fabrication, ETH Zurich, photos Callaghan Walsh

Norman Hack and Willi Viktor Lauer

MOULD

ROBOTICALLY
FABRICATED
SPATIAL MESHES
AS REINFORCED
CONCRETE
FORMWORK

If robots are to be employed in the construction of high-rise structures, logistics and material systems have to be entirely rethought to cater for both their abilities and limitations: robots have limited loading capacity, but greater dexterity than other on-site, automated construction techniques. Here, **Norman Hack** of ETH Zurich and **Willi Viktor Lauer** of the Future Cities Laboratory (FCL) at the Singapore-ETH Centre for Global Environmental Sustainability (SEC) describe the research they have undertaken to develop 'mesh-mould' as an alternative to conventional formwork.

Digital architecture in the 1990s was predominantly concerned with new computer-aided design strategies and was often criticised for neglecting issues of materialisation and construction.[1] The gap between what it is digitally possible to design and what is physically feasible to build narrowed when throughout the early 2000s CNC machines became more commonly available and eventually enabled designers and architects to bring their designs back from the virtual medium into the physical world. These machines offered unprecedented freedom for the fabrication of geometrically complex parts and intricate surfaces. However, they were largely limited to subtractive processes and prefabrication. With the introduction of industrial robots to architectural research,[2] the scope of digitally controllable fabrication processes widened dramatically. Unlike with most specialised machines, such as CNC gantry mills, the scope of the industrial robot is not defined and limited by its kinematics and offers an opportunity not only to customise the machined parts, but beyond that the entire fabrication process.

The generic, versatile and anthropomorphic nature of robots has inspired architecture students and researchers to equip these machines with almost every conceivable tool for gluing, melting, drilling, winding, cutting, pouring or painting. Even though, in the domain of architecture, robotic fabrication is still a new field of research, a remarkable amount of small yet sophisticated architectural structures have already been built and have impressively demonstrated the flexibility of such robots. Displaying a high degree of spatial and structural differentiation, these prototypical designs already hint at the potential for the application at a larger scale of load-bearing structures. However, until now applications in construction at the large scale have barely been investigated.

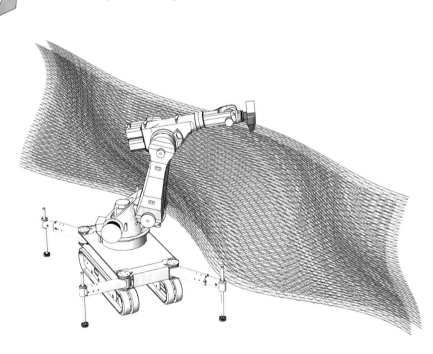

Fabio Gramazio and Matthias Kohler, Mesh-Mould, Architecture and Digital Fabrication, ETH Zurich, 2013

this page and previous spread: Over 60 per cent of the costs of a concrete structure are due to the labour-intensive construction of formwork, and bending and placing of reinforcement accounts for another significant share. Mesh-Mould proposes the unification of these two systems into one combined formwork-reinforcement system. The 3D mesh structures are robotically fabricated in an additive, waste-free manner, providing increased geometric complexity without raising the costs.

Conventional Construction Versus Robotic Fabrication

Embedded within the broad scope of investigations at the Future Cities Laboratory at the Singapore-ETH Centre for Global Environmental Sustainability (SEC), the Mesh-Mould research[3] focuses specifically on the mechanisms and the constructive processes of mass-produced architecture in Asia. In Singapore, over 80 per cent of the building stock consists of high rises, with the biggest share erected as prefabricated large panel system constructions. This dominant technique, driven by an economy of scale, becomes efficient only if the same elements are repeated over and over again. In a structural sense this practice results in an unsustainable over-dimensioning of a large number of structural members that have to be designed for the worst load case. In the bigger picture, the inherent limitations of such systems to incorporate variation do not allow programmatic or spatial differentiation, which ultimately results in an equally unsustainable oversimplified, mono-functional and monotonous urban fabric.

A shift away from this outdated 'one size fits all' ideology towards a new fabrication paradigm based on flexible, robotic in-situ fabrication could promote an alternative tectonic that encourages variation and differentiation instead of being bound to geometric simplification, standardisation and repetition. The feasibility of non-standard fabrication, as demonstrated numerously at the small scale, leaves little doubt over its potential at the construction scale, yet the leap into bigger dimensions carries various challenges and raises new questions of its ability to scale.

The Dilemma with Robots

If standard six-axis robots, applied in the manufacturing industry for years and a reliable and relatively cheap off-the-shelf technology, are to have applications in construction at the larger scale, a reconsideration of known material processes is required.

The construction logistics and the material system have to be designed to accommodate the specific abilities and limitations of robots. For instance, on the one hand the small size of robots is a limitation; on the other, it holds a significant potential. Their payload limitation does not allow for the handling of materials in the sizes and weights they are commonly used in construction. However, their small dimensions and low mass do make them suitable for application directly on the building site. Mounted on a mobile platform and equipped with sensors, multiple robots can navigate the building site, react dynamically to the tolerances and changing conditions of this complex environment, and fabricate the structure directly in situ[4] (see also Volker Helm's 'In-Situ Fabrication' on pp 100–107 of this issue). This conceptual approach differs greatly from gantry machines, which either have to be scaled beyond the size of the building footprint, as in the case of the Japanese construction automation systems (see Thomas Bock and Sike Langenberg's 'Changing Building Sites' on pp 88–99 of this issue), or, alternatively, the building elements have to be scaled down to match the envelope of the machine. This in return raises questions of segmentation, transport and assembly logistics on site.

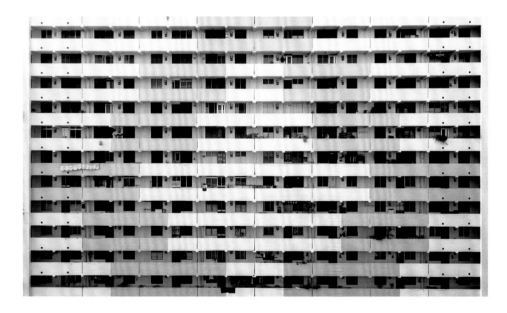

Housing and Development Board, Housing complex, Singapore, 2013

Due to the high fabrication costs of formwork, most conventional prefabricated concrete structures are based on the repetition of a few geometrically simplified elements. This 'one size fits all' approach does not only dramatically limit the freedom of design, but is equally unsustainable since far more material is used than is structurally needed. The robotic in-situ extrusion of formwork and reinforcement, in contrast, allows for locally differentiating the structure, using material only where it is needed, and thus liberating construction from the restraints of repetition.

Consideration of size, weight and payload has led to the conclusion that in order to be effective and competitive, an appropriate construction system for standard industrial robots needs to be lightweight, built from comparably small elements, and allow parallelisation and collaboration with other robots, humans and conventional construction machines. However, most importantly, the application of robots is feasible only if it generates a value-adding effect. The centralised, fully automated Japanese construction systems failed to do so, as they merely tried to automate existing processes, placing the emphasis on the elimination of human labour from the building site. They did not manage to capitalise on the real potential of machinic automation, they underestimated the complexity of the building process, and overlooked the fact that for many jobs humans are not only needed, but are also more efficient. Hence, different construction processes need to be developed to specifically address the strengths of robots in order that they can be applied where they actually outperform humans and conventional construction tools.

Division of Mass and Information

At first glance, the requirements for a robotic fabrication system seem to be somehow contradicting. Most common building materials for large-scale constructions are all but small and light. However, shifting the focus away from processing the entire mass of a building by robots, towards using them only for its geometric definition, solves the issue of limited payloads and simultaneously allows the benefits of their dexterity to be maximised. In this respect, concrete, the most commonly used construction material, has the inherent properties to be separable into its heavy structural mass and light, shape-defining formwork; in other words, it permits the separation of mass and information. In a future scenario collaboration between conventional tools, humans and robots, standard concrete pumps would transfer the actual structural mass, while the robot could unlock the inherent potential of concrete to take any desired shape by building complex formwork in high resolution. Due to the usually high costs of labour involved in the fabrication of custom formwork, this potential all too often remains inactivated.

The desire to fully explore the malleable potential of concrete has a long-standing tradition, beginning in the 20th century. Le Corbusier and, most notably, Pier Luigi Nervi, built complex curved concrete structures using manually assembled formwork. More recently, the increased geometric complexity enabled through digital design tools on one hand, and the technical possibility to mill Styrofoam inlays directly from design data on the other, have led to the emergence of a technique that combines custom inlays with standard formwork, which is state of the art and economically viable for a limited range of building typologies and budgets.

Danny Hill/Forma-Tech, Leaking Formwork, Victoria, Australia, 2010

The corrugated plastic panels are clipped together on site, holding in place the vertical and horizontal steel reinforcement. A fairly fluid concrete is tuned to entirely flow around the steel reinforcement and the plastic panels, but still needs to be sufficiently viscous to not completely leak out of the perforations. Since the operator can see through the perforated plastic panels during the concrete pouring process, he can subsequently densify the cavities; hence there is usually no need to vibrate the concrete.

However, in the past decade, academia and industry have made large efforts to eliminate the need of formwork entirely. Even though there are some differences, these approaches are best being generalised as layer-based 3D printing of cementitious, or concrete-like, materials. And though these developments offer hope regarding the potential to build freeform surfaces entirely waste-free, there are certain limitations and difficulties of such systems. For example, the hydration process of cementitious materials is difficult to control, and has a profound impact on the bonding behaviour among the layers. In the case of wrong timing, these do not sufficiently connect, which consequently downgrades an otherwise isotropic material to an anisotropic one, and thus limits the constructive capacities of the material. Furthermore, layer height, surface resolution and printing time are closely correlated parameters: in order to achieve a smooth surface, the layer height needs to be sufficiently small, which through every layer bisection cubically increases fabrication time. These issues have so far not been convincingly resolved and are reasons why recent research has shifted away from printing the actual concrete structure itself.

Mesh-Mould Combined Formwork and Reinforcement System for Concrete

Based on these findings, for the Mesh-Mould research project the decision was taken to concentrate on the fabrication of the formwork using an interesting technology known as 'leaking formwork'.[5] Here, concrete is poured into a corrugated formwork that is built up from perforated flat plastic panels and enables the erection of straight and single curved walls. The concrete then protrudes through the perforations and covers the formwork. In a subsequent step the surfaces are manually trowelled, leaving behind a smooth concrete surface.

Against this backdrop, polymers were freely extruded in 3D space, precisely controlled by the robot, in order to create the required meshes and liberate the formwork from geometric constraints. The use of thermoplastic polymers, such as used in conventional 3D printers, permits precise control over the material's hardening behaviour. Pinpoint cooling during the extrusion process, for example, gives such a high level of control that free spatial extrusions become possible and, consequently, the 'knitting' of structures freely in space.

Fabio Gramazio and Matthias Kohler, Mesh-Mould, close-up of extrusion process, Architecture and Digital Fabrication, ETH Zurich, 2013

This second generation of extruders is based on a combination of custom-machined parts, off-the-shelf 3D printer components and an additional air-cooling system. Pressurised air is directed at the extruder tip, so that the material can harden the moment it is extruded. Air-cooling, feed rate, extrusion rate and designated motion stops need to be carefully calibrated in order to facilitate a spatial extrusion and to minimise material-dependent tolerances.

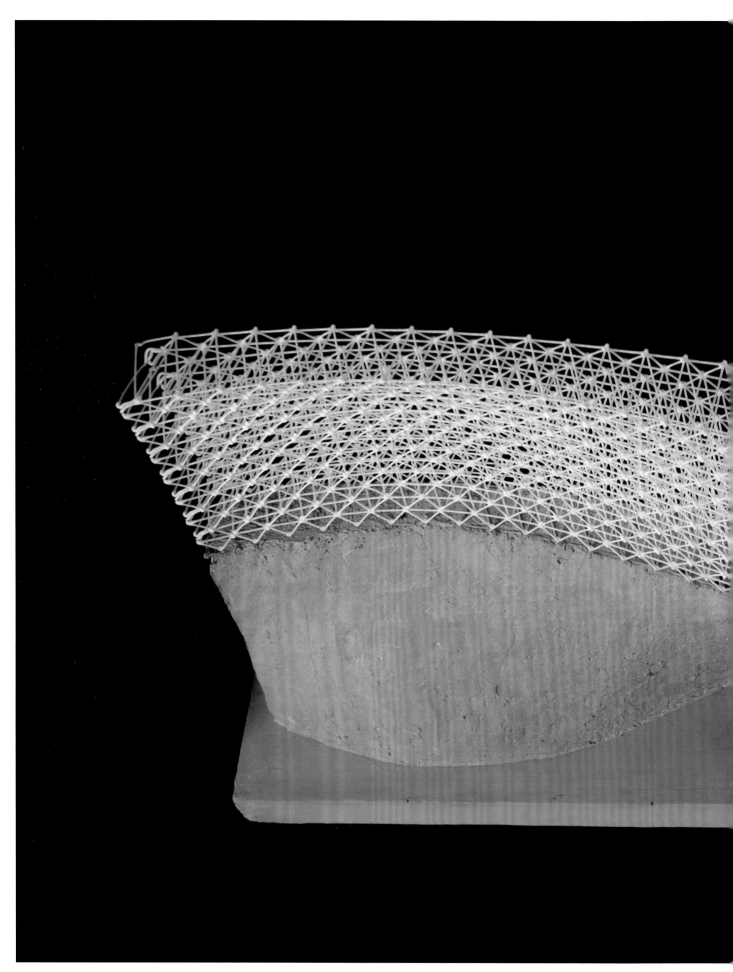

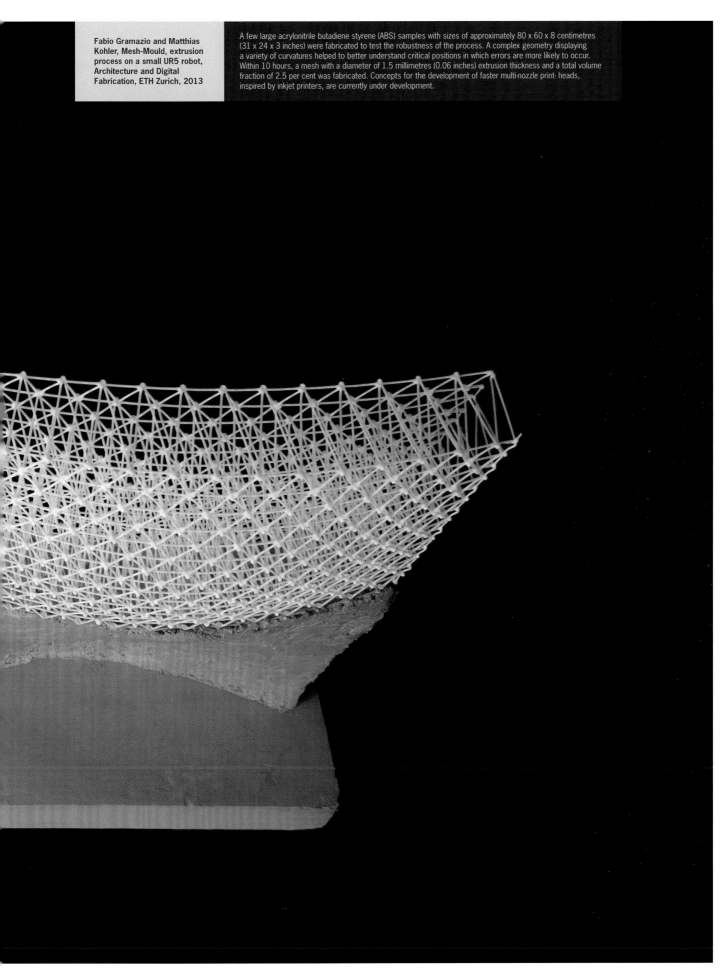

Fabio Gramazio and Matthias Kohler, Mesh-Mould, extrusion process on a small UR5 robot, Architecture and Digital Fabrication, ETH Zurich, 2013

A few large acrylonitrile butadiene styrene (ABS) samples with sizes of approximately 80 x 60 x 8 centimetres (31 x 24 x 3 inches) were fabricated to test the robustness of the process. A complex geometry displaying a variety of curvatures helped to better understand critical positions in which errors are more likely to occur. Within 10 hours, a mesh with a diameter of 1.5 millimetres (0.06 inches) extrusion thickness and a total volume fraction of 2.5 per cent was fabricated. Concepts for the development of faster multi-nozzle print-heads, inspired by inkjet printers, are currently under development.

This conceptual change, from layer-based material deposition to spatial extrusion, has several noteworthy implications. Whereas the former remains generic, mostly for the reproduction or representation of form, the controlled spatial extrusion becomes specific to the construction process. Firstly, the fabrication time is significantly reduced and becomes feasible for application in construction at the large scale. Furthermore, the mesh densities are generated according to the various forces that act on the structure. This applies for the static case after the concrete has cured, but also for dynamic loading during the concrete pouring process itself. Most interestingly, the alignment of the extrusion in accordance with the forces has the potential for co-extruding a strong filament such as carbon, glass, bamboo or basalt. This addition enables the structural activation of the mesh, making it accountable for high-tensile forces and ultimately replacing the entire conventional steel reinforcement. The robotic fabrication of freeform meshes not only allows for local adaptation towards various parameters like stresses and curvatures, but additionally enables the integration of cavities for lighter porous structures. Internal branch-like ducts for an enhanced concrete flow can be integrated directly in the algorithmic generation of the meshes and guarantee an optimal distribution of concrete within the structure.

Beyond these functional advantages, varying densities of the meshes can be used to generate advanced material effects, for example by keeping the concrete from reaching all the parts of the mesh, or inversely by making some parts of the meshes too large to hold back the concrete. Such material effects are specific to this system and could not be achieved by means of conventional mould-based formwork.

Potential for Design, Planning and Construction

The collaborative, distributed manner in which the mesh-fabricating mobile robots work together on the building site is scalable to various project sizes by parallelisation. By simply adjusting the number of collaborating agents, Mould-Mesh stays flexible and versatile, both in terms of design freedom and required infrastructural investment.

A typical concrete process involves a long and sequential chain of tasks from the prefabrication of formwork, transportation, site logistics, bending and placing of reinforcement bars, installation of formwork, concreting, disassembly and cleaning of formwork, and finally surface finishing. As the robots directly extrude the reinforcing formwork in-situ, several of these crafts and professions involved can be folded into one, allowing a higher product complexity while simplifying the process itself.

Under the consideration of Asia's incessant building activity and the sheer amount of buildings to be constructed in the near future, it is becoming increasingly important to develop sustainable construction systems that are cost sensitive, material efficient, and that provide for substantial architectural variation and programmatic differentiation. The unification of the two conventionally separate requirements of concrete – the reinforcement and the formwork – into one single robotic fabrication process, can produce an additive and waste-free, material-efficient and geometrically unconstrained method of fabricating complex non-standard concrete constructions. ᴧ

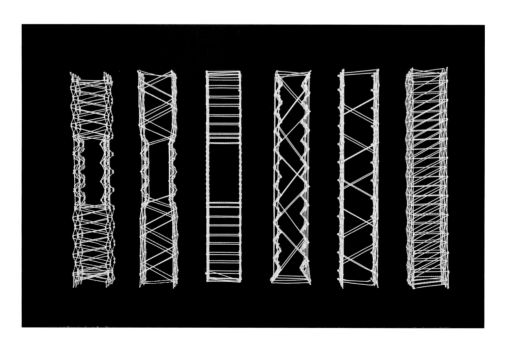

Fabio Gramazio and Matthias Kohler, Differentiated Mesh Patterns, Architecture and Digital Fabrication, ETH Zurich, 2013

A series of differentiated mesh patterns has been tested and evaluated according to various partially competing parameters. Structural stability, concrete flow and distribution within the mesh, fabrication time and material consumption can all be improved by various means of geometric differentiation. An improved concrete flow, for example, can be achieved by introducing ducts or voids, whereas increased structural stability is accomplished by folding the outer perimeters. The performative requirements for the meshes vary locally and the meshes can adapt accordingly.

Fabio Gramazio and Matthias Kohler, Mesh-Mould, tall prototype extruded with a mobile robot, Architecture and Digital Fabrication, ETH Zurich, 2013

As a prototypical setup for a process that is based on collaborative multi-agent in-situ fabrication, larger-scale experiments are carried out on an ABB robot mounted on a mobile platform. The sample was extruded to a height of 1.8 metres (5 feet 11 inches) and, apart from slight instabilities caused by the slenderness of the sample, has not displayed any major inaccuracies. Fabricated in 30 hours with an extrusion thickness of 2.5 millimetres (0.1 inches), the final sample weighs about 3 kilogrammes (7 pounds).

Notes
1. Antoine Picon, 'Architecture and the Virtual: Towards a New Materiality', in A Krista Sykes, *Constructing a New Agenda: Architectural Theory 1993–2009*, Princeton Architectural Press (New York), 2010, pp 270–89.
2. See, for example, The Programmed Wall, Architecture and Digital Fabrication, ETH Zurich, 2006: www.dfab.arch.ethz.ch/web/d/lehre/81.html.
3. Mesh-Mould is a PhD research project co-founded by speciality chemicals company SIKA AG, Switzerland, and conducted by Norman Hack.
4. Volker Helm, Selen Ercan, Fabio Gramazio and Matthias Kohler 'Mobile Robotic Fabrication on Construction Sites: dimRob', *Intelligent Robots and Systems (IROS), 2012 IEEE/RSJ International Conference on Intelligent Robots and Systems*, Vilamoura, Portugal, 2012, pp 4335–41: www.gramaziokohler.com/data/publikationen/969.pdf.
5. See http://www.formatechinternational.com/how-it-works/.

Text © 2014 John Wiley & Sons Ltd. Images: pp 44-7, 49-53 © Gramazio & Kohler, Architecture and Digital Fabrication, ETH Zurich; p 48 © Campbell McLennan

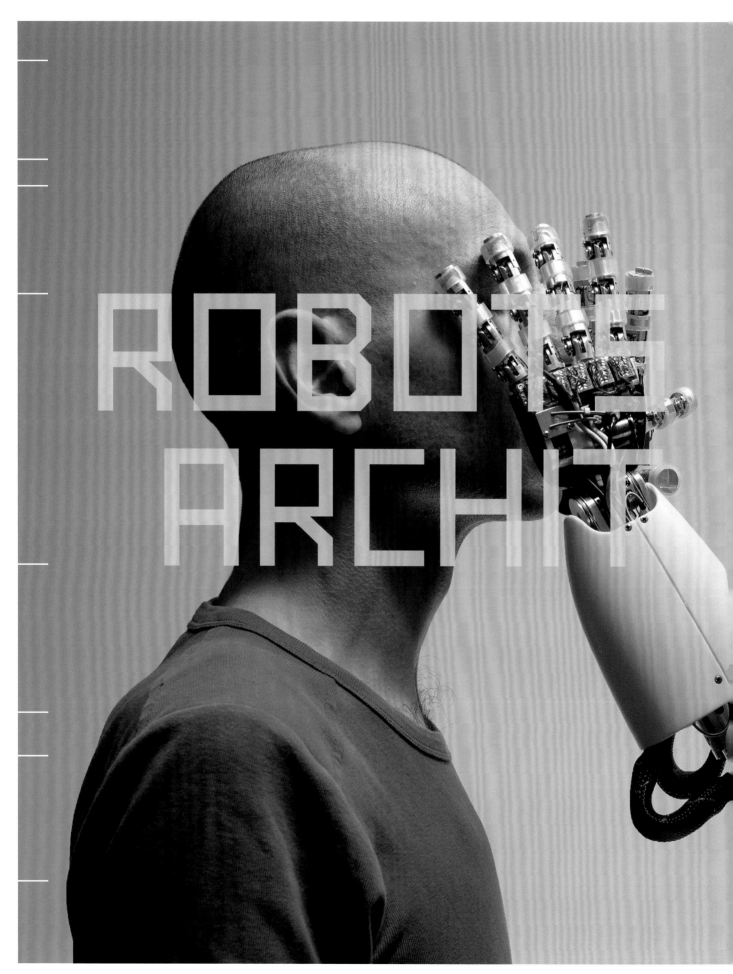

Antoine Picon

AND
LECTURE
EXPERIMENTS
FICTION
EPISTEMOLOGY

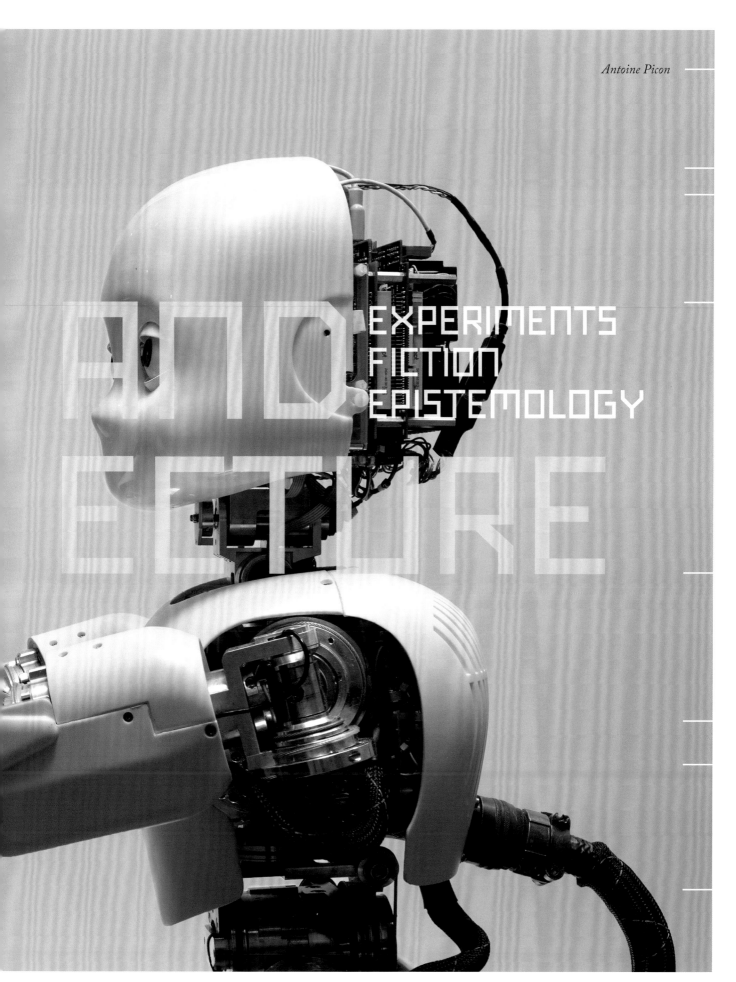

Albert Robida, *Leaving the Opera in Year 2000 (La Sortie de l'Opéra en l'An 2000)*, 1882
Gramazio & Kohler's ballet of flying machines for their Vertical Village is actually rooted in a long tradition of science-fiction images of future aerial urban life. This one by French illustrator Albert Robida represents fashionable society leaving the Paris opera at night after a show.

How ready are we to receive robots on our building sites? **Antoine Picon**, G Ware Travelstead Professor of the History of Architecture and Technology at Harvard Graduate School of Design (GSD), highlights the mixed cultural reception that robotics in architecture has received, veering from techno-utopianism to techno-pessimism. Is the greatest value in robotics for architecture in fact contained within the discipline, residing in the way that it forces architects to think differently, shifting their mental landscape and making them design truly three-dimensional space? Are we, though, in danger of neglecting to explore and re-imagine the fundamental relationships between men, designers and workers, and machines, computers and robots?

There are two very different but prevailing attitudes to technology, in regards to its capacity to innovate and make a difference. The first of these two attitudes is techno-utopianism. Innovators tend to share with entrepreneurs an unbridled optimism. You cannot be an entrepreneur if you are not confident that things will turn out for the better. Beyond a merely positive outlook, techno-utopianism is generally typified by a belief that innovation has an immediate and beneficial impact on society. There are numerous examples of this kind of approach. Richard Buckminster Fuller comes immediately to mind, with his unflagging endorsement of technological progress that he saw as a means to redesign society. According to Fuller, the Dymaxion house, car, and prefabricated bathroom that he developed in the 1920s and 1930s were not only ingenious devices meant to revolutionise the building industry, transportation and everyday life; they were also intended to pave the way for a radically different future in which men would roam free on the surface of the globe, live everywhere and fully take advantage of their intellectual capacities.[1] In contrast to techno-utopianism, the opposing view, techno-pessimism, is not that prevalent. It is largely limited to domains like literature and philosophy. You have only to think of the condemnation of railways by so many 19th-century writers, or the distrust expressed by 20th-century phenomenologists, such as Martin Heidegger, in the power of technology to improve the human condition. A more nuanced assessment of the real impact of technological innovation on the structures of production and society is far more common. Wary of discourses, which they often perceive as crude simplifications, historians have specialised in a more subtle and perceptive take on innovation. In his book *The Shock of the Old: Technology and Global History Since 1900*, British historian David Edgerton has tried, for instance, to reassess the respective roles of traditional techniques and highly visible breakthroughs like nuclear power in our contemporary world. His line of argument is that everyday techniques have been more instrumental in shaping what lies under our eyes than more widely and highly acclaimed innovations.[2]

Robots and Fiction

In the last few decades, the digital and its impact on architecture and construction have given birth to a wide array of discourses and statements that can generally be categorised as either techno-utopianism or techno-pessimism. While vibrant praise for digital technologies with their potential for positive change has become ubiquitous, more nuanced assessments of any ensuing transformations have also been produced. Recently, digital fabrication in architecture has become the focus of many of these unabashed eulogies and critical evaluations. Are we facing a revolution comparable in scope with the invention of printing in the Renaissance? Or does it represent a more finite evolution? A more tentative hypothesis is after all tenable, given the gap between the still highly experimental character of digital technologies and the predominance of traditional mass-production techniques, if not their further diffusion, in countries like China.

This dual regime is even more pronounced when automating construction through the use of robots. On the one hand, following the pioneering experiments of Fabio Gramazio and Matthias Kohler at ETH Zurich, where they share the Chair of Architecture and Digital Fabrication, robots appear as a key element of future architectural development. Many schools of architecture are now equipped with robotic arms that are used for structural investigations as well as for research on surfacing and patterning. On the other hand, it is easy to measure the limitations of the use of robots on ordinary construction sites, beginning with the innumerable problems linked to security and maintenance that they raise. In addition, they still possess an unassailable association with science fiction; a connotation that is especially marked in the experiments involving flying robots led by Gramazio and Kohler.[3] The aerial ballet of their insect-like machines assembling blocks with superhuman precision is evocative of aerial traffic patterns in cities of the future where flying vehicles have become common. Already present in Albert Robida's late 19th- and early 20th-century engravings and novels, this vision of the urban future would later become a standard of science-fiction movies, from Fritz Lang's *Metropolis* (1927) to Ridley Scott's *Blade Runner* (1982).

The science-fiction overtones of robotics may well serve as a convenient transition towards one of the key roles of robots these days, namely their supporting part in a narrative regarding the future of the architectural discipline and the rising importance of automated fabrication. This type of narrative dimension is by no means a recent phenomenon. The various attempts to industrialise building activity throughout the 20th century were intimately related to a grand narrative regarding the necessity to adapt architecture to the age of the machine.[4] One of its interests was to enable its proponents to occupy a middle ground between overt and thus disputable techno-optimism, the fictional mode suspending questions of immediate feasibility, and the opposite sceptical attitude. Experiments extended into fiction as a way to suggest, without unnecessary heaviness, that they could lead to widespread change.

Utopian Perspectives

Despite its attempt to distance itself from techno-utopianism, the narrative of the industrialisation of construction remained permeated by utopian concerns such as the desire to reconcile nature and technology, the project to free man of unnecessarily harsh work, or the ambition to enable man to live everywhere on the planet, the latter being especially present in Buckminster Fuller's approach to industrialisation. In many respects, the advent of robotics in architecture is marked by the emergence of a similar type of narrative. On a certain number of points, this new narrative appears as the direct

Gramazio & Kohler and Raffaello D'Andrea with ETH Zurich, *Flight Assembled Architecture*, FRAC Centre, Orlèans, France, 2012

Flying robots assembling the Vertical Village, a megastructural project designed by Gramazio & Kohler. The design of the Vertical Village is intimately dependent on the project to assemble it using flying robots.

inheritor of the industrial one. Like its forerunner, it is permeated by utopian themes, some relatively traditional, others without equivalent in the history of industrial ideals. Whereas the project to relieve man of painful tasks is by no means original, the quest for a new immediacy based on computation between the designer's mind and the built reality is without precedent.

Throughout the 20th century, machines used to prefabricate or customize, and to assemble parts, had been interpreted as tools radically distinct from the mind that put them in motion. Equations and flows of data seem to constitute, in contrast, a fluid milieu that tends to unite the human brain and its mechanical extensions. This new intimacy could be described as the advent of a cyborg designer whose intentions are materialised through the action of powerful artificial arms. But this perspective may be misleading insofar that the best way to envisage what robots do is not necessarily to consider them as extensions of the human mind and body. For they do not exactly replace human arms and hands; they follow principles of their own, often different from the rules that govern human productive gestures. Coupled with the readiness with which they obey the designer's instructions, such difference increases their epistemic potential, as we shall see towards the end of this article.

Fabio Gramazio and Matthias Kohler, The Sequential Wall, Architecture and Digital Fabrication, ETH Zurich, 2009
A robotic arm in use. All over the world, robotic arms have become standard equipment in design schools that can afford them. Here, the robot assembles a wall integrating a series of functional requirements as generative parameters of the design.

Why not imagine a unified design and fabrication process based on a series of conversations between men, designers and workers, and machines, computers and robots?

Besides the new immediacy between the mind and the built reality, robotics announces the possibility to radically overcome the constraint of large series that had hampered so many former attempts to industrialise construction. The utopian perspective of a world of 'makers', to use Chris Anderson's term, becomes almost inescapable.[5] In this world, prototyping and small-scale production of sophisticated components would advantageously compete with repetition and mass production.

The enthralling spectacle of a harmonious ballet of productive forces and machines, of a streamlined universe of efficiency and elegance, appears also inescapable. On this ground again, the robotic fabrication narrative is not entirely original. Modernist industrialisation had already dreamt of streamlining projection. The possible autonomy of robots could constitute nevertheless a striking point of departure from the vision developed throughout the 20th century by the advocates of rationalisation. In science-fiction novels and films, robots appear sometimes as obedient slaves, sometimes as dangerously independent life forms. Recent and spectacular advances in intelligent automation are, day after day, making the latter alternative more plausible. Robotic fabrication may confront us for the first time directly with the need to cooperate with our technological auxiliaries rather than simply use them.

The human workforce seems to be so far missing from this narrative, as if an exclusive dialogue between designers and robots were the only development worth exploring. More generally, discourses on digital fabrication in architecture often indulge in a strange kind of Ruskinianism. In their neo-Ruskinian perspective, designers tend to occupy the place formerly devoted to craftsmen, that of inspired artisans shaping the world with their hands – digitally augmented hands that is. Understanding the new immediacy between mind and matter in such a way is doubly misleading. It forgets both the persistent need for a human workforce to step into the gap left by robots' shortcomings and their potential otherness that forbids considering them as mere extensions of the designer's hand.

From Fiction to Conversation

Will the fiction one day become reality? Although the grand industrial narrative of the 20th century never came fully to fruition, its legacy was considerable, from new materials like plastics to key techniques of dry assemblage. The robotics narrative will probably have equally enduring effects on the built environment, and this is all the more feasible given that the digital age is marked by the multiplication of auto-realising fictions. Before the multiple connected devices that surround us were realised, ubiquitous computing was one of these fictions.[6]

But the impact of current robotic fabrication experiments on architecture may extend beyond improved materials and innovative techniques. Returning again to modernist industrialisation, it is striking to observe how it influenced the fundamentals of the architectural discipline by redefining both what design was about and the effects and affects it was supposed to produce. To use Walter Gropius's characterisation, industrialisation was not only meant to

Fabio Gramazio and Matthias Kohler, *Structural Oscillations*, Architecture and Digital Fabrication, ETH Zurich, Venice Architecture Biennale, 2008

Structural Oscillations, an installation at the 11th Venice Architecture Biennale, showcases the capacity of algorithmic design and robotic assemblage to create a stable overall configuration from individual brick layers that could be tipped over by visitors if left alone.

produce buildings differently; it was to represent a 'purifying agency' liberating architecture from outmoded technological as well as aesthetic values.[7] In other words, it played an epistemological role and forced designers to think differently.

Such might be the most important consequence of the introduction of robots in architecture, at least for now. The large success met by the experiments conducted by Gramazio and Kohler at ETH Zurich has to do with the clarity with which they have outlined this epistemological dimension. This led them to propose in their first book the notion of a new digital materiality redefining the way designers address the fundamental problems of their discipline.[8] In addition to the characterisations of this new materiality they proposed – like the simultaneous rise of sensuality and of programming, the growing importance of variation and multiplicity and the key role of processes, or rather in a slightly different perspective – the following observations can be made.

First, robots do force designers to think in a truly three-dimensional space in which there are no longer any privileged directions. Moreover, they remind us of the foundational character of rotation in the analysis of motion. The movements of our body are themselves based on the various rotations of our members. But we have for a very long time forgotten this simple fact and used rectilinear motion as the standard spatial operation, from mechanics to design. Modernist industrialisation itself relied heavily on repetition by rectilinear translational motion. In short, robots introduce us to a profoundly different geometric world. Even if gravity will continue to make us distinguish between the vertical and the horizontal, while translational motion will remain an important feature of mechanics, the mental landscape of design is about to shift.

Second, we have known for a while that the emergent digital materiality was based on the association of notions that used to be antagonistic: the physical and the electronic, the sensory and the computational, the concrete and the abstract. Robots teach us that there is also a thinner and thinner line between objects and processes, as well as between stability and instability. As Gramazio and Kohler have once again demonstrated with experiments like their 2008 *Structural Oscillations* installation, robots enable designers to play at the very frontier that used to separate stable assemblages from unstable combinations.

Third, robotic fabrication in architecture induces a change in efficiency as well as in the beauty that is linked to it. There used to be a superiority attached to the simple and the scarce over the complex and the multiple, a superiority rooted in technological reminiscences. According to scholastic philosophy, simplicity and unity represented fundamental attributes of the divinity. Digital technology and robots are radically challenging such a perception. The complex and the multiple appear more and more as the natural condition from which designers should start.

What remains to be explored is the potential of the machine to emancipate itself, at least partially, from the instructions of the designer in order to appear as a significant other in the conception of the project. Whereas architects like Cedric Price had tried to follow this path in the early 1970s[9] – probably too early in terms of technological feasibility – the possibility of such otherness is often lacking today from the speculations and experiments regarding digital design, just like the role devoted to the workforce is generally minimised. Everything happens as if computers and robots were to remain forever obedient slaves, while the role of workers is steadily diminishing.[10] Overcoming this strange shortsightedness could very well represent the next step in the investigations regarding the use of robots in architecture. Why not imagine a unified design and fabrication process based on a series of conversations between men, designers and workers, and machines, computers and robots? A truly different architecture could rise from such extended conversations. ⌂

Notes
1. Michael Hays and Dana A Miller (eds), *Buckminster Fuller: Starting with the Universe*, Whitney Museum and Yale University Press (New York and New Haven, CT), 2008.
2. David Edgerton, *The Shock of the Old: Technology and Global History Since 1900*, Oxford University Press (Oxford), 2007.
3. Fabio Gramazio, Matthias Kohler and Raffaello D'Andrea (eds), *Flight Assembled Architecture*, Editions HYX (Orléans), 2013.
4. See Antoine Picon, 'The History and Challenges of Industrialised Buildings in the 20th Century', in Franz Graf and Yves Delemontey (eds), *Understanding and Conserving Industrialised and Prefabricated Architecture*, Presses Polytechniques et Universitaires Romandes (Lausanne), 2012, pp 63–70.
5. Chris Anderson, *Makers: The New Industrial Revolution*, Crown Publishing Group (New York), 2012.
6. Paul Dourish and Genevieve Bell, *Divining a Digital Future: Mess and Mythology in Ubiquitous Computing*, MIT Press (Cambridge, MA), 2011.
7. Walter Gropius, *The New Architecture and the Bauhaus*, Museum of Modern Art and Faber and Faber (New York and London), 1936, p 19.
8. Fabio Gramazio and Matthias Kohler, *Digital Materiality in Architecture*, Lars Müller Publishers (Baden), 2008. On the notion of digital materiality see also Antoine Picon, 'Architecture and the Virtual: Towards a New Materiality?', *PRAXIS*, 6, 2004, pp 114–21.
9. See Antoine Picon, *Digital Culture in Architecture: An Introduction for the Design Professions*, Birkhäuser (Basle and Boston, MA), 2010, p 37.
10. Kostas Terzidis remains an exception in that respect. See his *Algorithmic Architecture*, Architectural Press (Burlington, VT), 2006.

Text © 2014 John Wiley & Sons Ltd. Images: pp 54-5 © Emile Loreaux/Picturetank; p 57 © FRAC Centre, photography François Lauginie; pp 58-9 © Gramazio & Kohler, Architecture and Digital Fabrication, ETH Zurich

ENTREPRENEURSHIP IN ARCHITECTURAL ROBOTICS

Hyperbody, RDM Vault, Rotterdam, 2012
The RDM Vault explores compression-only structures assembled from extruded polystyrene foam components. The project was designed and built in three weeks.

THE SIMULTANEITY OF CRAFT, ECONOMICS AND DESIGN

Jelle Feringa

What do Odico Formwork Robotics, RoboFold, Machineous, ROB Technologies and GREYSHED share in common? They are all architectural robotics startups. **Jelle Feringa**, Chief Technology Officer at Odico, places the phenomenon of the architectural robotics entrepreneur in a historical and cultural context while highlighting the very practical role startups are poised to play in bridging the gap between academic research and industry, by providing the building industry with much needed new software tools.

If I was to realize new buildings I should have to have new technique. I should have to so design buildings that they would not only be appropriate to materials but design them so the machine that would have to make them could make them surpassingly well.
— Frank Lloyd Wright, 1932[1]

What are the prospects of entrepreneurship in architectural robotics?

Do forward-looking, entrepreneurial architects share kinship with the Renaissance architect Filippo Brunelleschi, who in the 1420s engineered a giant, three-speed, reversing ox-driven hoist, enabling him to span the massive crossing of Florence's cathedral with a dome? Is Jean Prouvé, an architect who ran a factory and who played a pivotal role in developing cutting-edge production technology and modular systems, a 20th-century analogy of how architects will operate in the 21st? Or will education tragically monopolise patronage of architectural robotics?

Sharing the technological platform of robotics with industry encourages the transfer of knowledge from academia to industry. Researchers share the parlance of industry – code interpreted by the robot controller – with dialects differing from ABB's RAPID, Kuka's KRL or Staübli's VAL3 robot language. Operating on a similar platform challenges where technological innovation stops and industrial adoption starts; industrial robotics allows new manufacturing technologies to gradually evolve from initial experiments to industrial processes.

The modest investment required to bootstrap a startup company,[2] given the cost of a new or refurbished robot seen in relation to its potential returns and production capacity, coincides both with a renaissance in manufacturing[3] and a need to renew the architectural profession.[4]

Startups

A number of promising startup companies have been surfacing over recent years, leveraging architectural robotics beyond mere conceptual merit and stepping into the industrial arena – where startups play a pivotal role. While the robot research work is creating considerable interest, without the filter of practical adoption – technology you can buy – expectations go unfulfilled.

This is why the startups presented over the following pages occupy an essential position in bridging the domains of academia and industry. Progress is affected by the cynicism sometimes observed in these domains, while institutionalism fuels the opportunity. The arguments are that from an academic or a computer science perspective there is little novelty. Inverse kinematics solvers and offline-programming software are sometimes considered to be 'known' problems.[5] Or from an industry point of view, the approaches developed in architectural robotics are far-fetched or perceived as too costly to develop, and as yet have an uncertain future in terms of their economic yield. The startups here challenge such long-proliferated misconceptions.

The lack of software tools suited to architectural processes might explain the relative and creative dilettantism in developing robotics software for architecture, developed by architects. However, the robotics entrepreneurs featured here focus on pushing novel approaches beyond their conceptual infancy, bridging the gap between academic merit and industrial adoption, providing architecture and the building industry with an important new set of tools. Traditional robot integrators – often coming from a background in the automotive industry – are yet to step into this domain. The fabrication processes developed in architectural robotics differ considerably from the repetitive routines of industrial automation. For example, industrial processes are often programmed in-situ, where a robot programmer is explicitly jogging and feeding the robot with instructions. Once instructed, a robot is able to repeat this procedure. This mechanical approach does not extend to architecture, where the umbilical cord to CAD geometry cannot be cut.

Recovering Lost Ground

This persisting CAD connection is why all of the startups presented here are so focused on software development – where the limits of where design stops *and* production starts are becoming increasingly intertwined. ROB Technologies and RoboFold develop both design and production tools. Are these companies forming a second wave after seeing the rise of the likes of designtoproduction, 1:One and CASE?

Rather than making the most of the array of conventional CNC fabrication methods,[6] these companies take up the role of introducing novel fabrication processes, ranging from stacking bricks to folding aluminium panels and cutting massive blocks of foam. By making the software and design tools available to architects and industry, the threshold for market adoption is lowered considerably.

Is it possible that, when an earlier generation of architects ceded greater responsibility for the realisation of buildings to engineering offices, architecture lost some of its professional authority? The success of design rationalisation offices, file-to-factory processes and architectural robotics suggests so. Are the aforementioned companies and the robotic startups stepping into this void and recovering lost ground? If anything, the discrepancy between what is practically and economically feasible, in comparison to the daily routine of building production, is enormous; a shared thread is found in the ambition of closing this rift.

An economical perspective brings about the promise of novel fabrication methods and architectural robotics. However oversimplistic it is to consider architecture as the difference between the raw building materials and the cost of their assembly, it is here where the difference is made.

Agriculture and construction are both the largest and least automated industries – where in the latter the cost of assembly essentially equates to the cost of labour. The economics of building have marginalised architectural ambitions, since craft is costly. However, as the cost of mechanised labour continues to drop dramatically, as automation and robotics become so ubiquitous, the moment has come to revitalise the architectural profession. Gramazio & Kohler's early Gantenbein vineyard facade (Fläsch, Switzerland, 2006) was seminal in this respect, simultaneously exploring the approach of automated bricklaying and its architectural potential. It is precisely this simultaneity of craft, economics and design that is so striking.

Design and Production

The BrickDesign tool developed by ROB Technologies (pp 72–3) further blurs the false schism between design and production. BrickDesign is a Rhino® plug-in encapsulating the expertise required to design and plan large assemblies of discrete elements – bricks. To bring ROB Technologies' building process of automated bricklaying in non-standard formations to the market, it was necessary to facilitate the means of design. A lack of proper software tools means it is virtually impossible to design such a number of discrete parts without falling back into traditional bonding rules, which would not exploit the potentialities of the robotic process. As one of the pioneering architectural robotics technologies, developed by architects for architects, BrickDesign is an important precedent of what is required before a fabrication concept can reach the levels necessary for application at the industrial scale. It has both democratised and opened up the approach to towards market adoption. This is a significant achievement, since traditionally industrial robot integrators have not been able to develop the required design tools, and as a result have only provided the required process automation.

The design and realisation of projects and prototypes built over the years by means of robotic bricklaying is an integral part of the push towards industrial readiness. For example, the client for a recent project – the brick facade of the Keller AG Headquarters (Ofenhalle, Pfungen, Zurich, 2012 – see page 72) – is in fact a central industrial partner of ROB Technologies, not only demonstrating confidence in the method developed, but also challenging the cliché of industry not being willing to commit to an investment in novel technology. Here is a partner that is pushing architects to reach new heights, both architecturally and in terms of technology, renewing the relationship between architect and industry. Regrettably, it is rare to see architects and their industrial partners changing the course of their industry in this way, hence the significant role startups such as ROB Technologies and the others featured here might play in progressing the field of architecture.

These startups are not just passively facilitating a construction method: an architectural ambition precedes an architectural technology. The recent Arum installation by Zaha Hadid Architects, RoboFold and Philippe Block (see pp 68–9), presented at the 'Common Ground' Venice Architecture Biennale in 2012, is insightful here. While a number of authors are credited, without RoboFold providing its own design software and fabrication process the constructive and aesthetic potential of the approach would not have been realised. The project was influenced by RoboFold's earlier developments, and as such this robotic startup company can be considered as both design *and* production partner. Their software tools aid in exploring the possibilities opened up by the robotic folding process – much like ROB Technologies' BrickDesign software allows exploration of the design space the manufacturing technique offers.

The question of economy, of developing an affordable approach to the fabrication of formwork, is addressed by Odico Formwork Robotics (pp 66–7). The manufacturing of sophisticated formwork accounts for more than three-quarters of the costs associated with the realisation of sophisticated geometry in concrete. By upscaling wire-cutting technology to a robot with a reach of over 25 metres (82 feet), the company can produce intricate polystyrene moulds for sophisticated concrete structures in a short timeframe and within moderate budgets that challenge existing approaches to formwork. It provide solutions for the scaling of fabrication processes to architectural dimensions, developing technology that is congruent with the dominant building method – casting concrete. By moving beyond the limited production capacity of a

> These startups are not just passively facilitating a construction method: an architectural ambition precedes an architectural technology.

CNC router, a cost-effective approach for the realisation of large sophisticated concrete structures becomes possible. Alongside its efforts for the building industry, Odico also produces formwork elements for a number of companies in the cleantech industry, such as Siemens Wind Power.

Recent explorations in stone-cutting technology by the Hyperbody research group at the Delft University of Technology underscore the interwoven relationship between manufacturing technology, economy and architecture.[7] Being able to process stone with the modest means of an industrial robot rapidly eludes the idea – an archaic notion – of building in massive stone elements. In addition, a salient realisation early on in the project is that the cost of marble has considerable scope. Since only up to 75 per cent of the stone quarried yields material that is of pristine quality, there is a large volume of second- and third-rate quality that is often ground down to a polishing agent used in toothpaste. The difference in quality is roughly based on the whiteness and homogeneity of the colour of the material, and translates into a factor of over 40 in the price per tonne. Given this economic bandwidth, an efficient way to shape the raw material into '*traits*'[8] suggests that building in stone may be a more viable option than is often considered.

This understanding has been one of the motivations driving the project, raising the question whether aspects of economy are an inherent element of design. Just as CNC and robotic technologies have made fabrication a more central design aspect, are such manufacturing technologies in turn bringing economic considerations into focus? ROB Technologies creates software, and RoboFold and Odico both develop software and produce building products. What will surface when the approach is extended to design?

Dutch designer Dirk Vander Kooij offers an interesting analogy. By mounting an extruder to an old FANUC robot, original and affordable chairs are produced by stacking contours layer by layer, much like an oversized 3D printer. What is interesting here is not only the quality of the end product, but that the designer is not relying on a third party to develop the process or produce the chairs, challenging the traditional boundaries of the profession, and certainly so from an economic point of view: 'If he is lucky, the designer gets 3% ex factory. The brand adds 300% and the shop doubles that again. It's ridiculous how little of the cut a designer gets. If we used digital tools and changed the way stores work, the ratio would be able to favor creativity and the craftsman.'[9]

Design and Build

It would be misleading to mistake architecture for industrial design; even at the scale of the house, it is over-simplistic to think of architecture as a product. This said, the raw production capacity of industrial robotics does bring 'design and build' approaches to construction into view. Are the startups in architectural robotics revisiting the idea of an architect with a method of design *and* the means of production? In 1996, Bernard Cache's company Objectile set up a factory utilising CNC milling machines. In 2000, architect Bill Massie built the Big Belt house, and more than a decade later companies like Facit Homes are revisiting the idea of the house as a product, where CNC is the enabling technology. Do these projects suggest a reconsideration of the early objectives of Modernism, to provide affordable and modern houses of architectural ambition?

To what extent the new-found vicinity of construction is desirable remains an open question. The emerging ecology of knowledge and new possibilities from which both the architect and contractor profit is good news for architecture. Combining architectural ambition with a sense of economical pragmatism, robotic entrepreneurship challenges preconceived ideas of what it is possible to realise given a building budget – an essentially architectural agenda is shared by these young companies.

Machineous is a fabrication and R&D company that has the most experience in this domain (pp 70–71). Since its inception in 2008, the company has sized up its operation to five robot stations, and recently moved to a large 1,800-square-metre (19,400-square-foot) facility in Los Angeles, completing the transition from a robotic artisanal workshop to an industrial operation. The company's rapid expansion is matched by its increased scaling, from the production of bespoke furniture, installations and public sculptures to the production of building elements ranging from stairs to window apertures and facade panels.

The crises of the American automotive industry between 2008 and 2010 made plenty of inexpensive but capable robots available at a fraction of the cost, and Machineous has adopted a number of these from a former Chrysler plant. The automotive industry's technological compost heap also fostered another startup: GREYSHED, a design-research collaborative focused on robotic fabrication within art, architecture and industrial design (see pp 74–5). GREYSHED's research has seen the integration of technologies such as augmented reality, gestural and sensor feedback in closely coupled design and fabrication processes. In its anti-institutional approach, experimentation and development of fabrication strategies are fuelled by the re-appropriation of affordable, off-the-shelf technological commodities ranging from smartphones and Kinects to chainsaws, cordless drills and refrigerator parts.

Given the current economic climate of the building industry, the momentum and market belief in the 'third industrial revolution', the impetus felt throughout the architecture, engineering and construction industry, the potential return on investments, and the low investment required to start developing architectural robotics, it is surprising that there are so few startups currently active. Are robotics and fabrication offering an apt bypassing of the maelstrom of architectural competitions, focusing on what is essential in architecture, or is there a derailing effect from taking up such a central role in the building process? Could it be Brunelleschi, the architect-engineer, versus Leon Battista Alberti, the intellectual architect, all over again? 𐤀

Notes
1. Frank Lloyd Wright, *Frank Lloyd Wright: An Autobiography*, Longmans, Green and Company (Toronto), 1932, p 149.
2. The tour ABB IRB 6400M94A robots that make up Hyperbody's robotics lab were acquired for the approximate cost of an entry-level 3D printer.
3. Paul Markillie, 'A Third Industrial Revolution', *The Economist*, 21 April 2012: www.economist.com/node/21552901.
4. Amanda Hurley, 'Available: Immediately. For architects, this time, it's personal: Mass layoffs at design firms and a tight job market bring the recession home', *Architect*, 10 February 2009: www.architectmagazine.com/architects/available-immediately.aspx.
5. In my opinion, mistakenly so. Only relatively recently have robots started to perform tasks such as milling and laser/plasma cutting. These tasks are highly demanding in terms of motion planning and optimisation, and require sophisticated, optimising inverse kinematics solvers. Such processes challenge the capability of existing motion-planning algorithms.
6. Such as laser, plasma and water-jet cutting, and milling.
7. Hyperbody is directed by Professor Kas Oosterhuis. Its robotics lab was initiated and is directed by Jelle Feringa.
8. A '*trait*' is a kind of drawing, usually drawn to scale, that contains the geometrical and constructive information needed to carry out the stone cutting for a specific work.
9. Gabrielle Kennedy, 'Joris Laarman's Experiments With Open Source Design', *Open Design Now: Why Design Cannot Remain Exclusive*, BIS Publishers, 2011, pp 200–8: http://opendesignnow.org/index.php/article/joris-laarmans-experiments-with-open-source-design-gabrielle-kennedy/.

Text © 2014 John Wiley & Sons Ltd. Images: pp 60-1 © Jelle Feringa

Asbjørn Søndergaard

ODICO FORMWORK ROBOTICS

When Odico Formwork Robotics was founded in April 2012, it was apparent that to bring the fabrication principles of architectural robotics beyond the laboratory and prototype project, it would be necessary to fully confront market conditions. The founding partners – two architects, one engineer and one investor – thus set out to develop a manufacturing concept capable of providing low-cost production of advanced formwork at industrial volumes of scale. They were convinced that the internal, comparative tests of the Robotic Hotwire Cutting (RHWC) technique they had developed through their architectural research in 2009, exhibited a greater production capacity than equivalently scaled CNC-milling processes.[1] When Odico began the commercialisation of this technology in 2012, it was an international first.

Currently hosting 15 employees in its Funen-based facilities in Denmark – six of whom are industrial ABB robots – Odico has a business model that revolves around two main activities: formwork production, and research and development (R&D) technology. Targeted at the concrete industry, the ongoing manufacturing provides advanced polystyrene moulds for architectural production. Replacing in a matter of minutes formwork previously requiring days of handcrafted labour, the factory production covers widespread, high-volume applications such as stairs and facade components, as well as bespoke, cultural artefacts in the form of sculptures and architectural structures. Complementing the manufacturing services provided for the construction industry, a significant part of Odico's production resources are occupied by architectural-scale industrial prototyping and manufacturing for the wind-turbine and wave-energy industries.

Alternating between physical production experience and architectural ambition, the second strand of the firm's business strategy – the R&D technology – facilitates the maturation of manufacturing concepts along with incubation of new technological ideas. The activities are focused in two primary areas: the provision of robotic cell design licences and fully automated, lights-out manufacturing processes. The hardware development laboratory is able to fully exploit the entirety of the factory space for 1:1 concept testing, while software development engages the sophistication of Odico's core enabling technology: PyRAPID, a pythonOCC-based standalone RHWC-CAM application. Written by a chief technology officer with a purely architectural background, it exemplifies Odico's founding spirit; that a technology innovation led by architectural aspiration can provide practical and economical industrial solutions that liberate, rather than delimit, architectural vision.

Odico Formwork Robotics hardware development lab, Odense, Funen, Denmark, 2013
A perpetual redefinition of tooling, configurations and functions facilitates the testing of experimental production setups where robots, end effectors and externalities interchange in a 1:1 sketching of new manufacturing concepts.

The R&D technology is nurtured through strong relations with academia. Within its first year of operation, Odico headed the three-year BladeRunner research project funded by the Danish National Advanced Technology Foundation. Conducted in collaboration with leading practice, university and industry partners, the project investigates the low-cost production of large-scale double-curved free-form geometries through the development of robotic hot-blade cutting technology. Complementing this is the two-year Opticut pilot project in which cost-efficient realisation of topology is being optimised, and concrete structures explored through ruled surface rationalisation and RHWC fabrication. Through the fabrication of a 20 x 3 x 5 metre (65 x 10 x 16 foot) prototype structure at Ärhus Bay, preliminary results indicate an approximate 24-fold increase in production capacity over comparable CNC techniques, leading to an 80 per cent reduction in machining time costs, thereby paving the way for realising greater architectural ambitions within moderate budgets.

Inspired by the ideas of Jean Prouvé, that to renew the profession architects must engage deeply in the design and development of construction technology itself, Odico continues to pursue the objective of introducing disruptive CAM technologies to the architectural industry. Tentatively exploring this notion in a wider tectonic context, the applicability of the stereotomic experiences from Odico's RHWC to robotic diamond wire cutting of marble in recent experiments, indicate that one measure of the robustness of a technological idea lies in its transferability across materials and platforms. ᗘ

Note
1. See Jelle Feringa, 'Investigations in Design and Fabrication at Hyperbody', and Per Dombernowsky and Asbjørn Søndergaard, 'Unikabeton Prototype', both in Ruairi Glynn and Bob Sheil (eds), *FABRICATE: Making Digital Architecture*, Riverside Architectural Press (Waterloo, ON), 2011, pp 56–61 and 98–105.

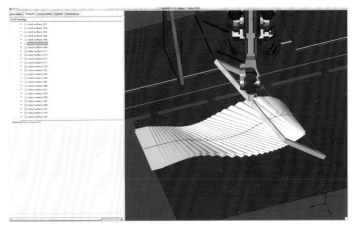

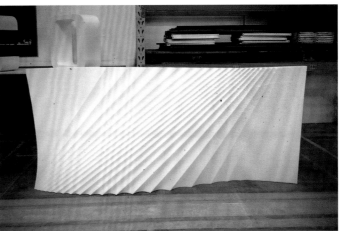

Odico Formwork Robotics, PyRAPID transliteration of ruled surface input geometry to an RHWC-cut polystyrene mould, 2013
Odico's PyRAPID inverse kinematics solver computes sequences of motion frames from the isocurvatures of incoming surfaces, outputting instructions in ABB-native RAPID code language.

Text © 2014 John Wiley & Sons Ltd. Images © Odico Formwork Robotics

Gregory Epps

ROBOFOLD AND ROBOTS.IO

RoboFold was founded by Gregory Epps in 2008, after using curved folding obsessively as a method of design for more than 17 years. The technology translates the intuitive process of folding paper by hand into an industrialised system designed to fold metal using multiple industrial robots. The company builds on the knowledge of how to work with the aesthetics and physics of sheet metal to achieve effective design solutions.

This novel approach demonstrates that innovation can reverse the ingrained concepts of conventional metal forming, a dichotomy that can be summarised through two opposing approaches. The conventional approach requires up-front financial commitment to the mould tooling used to form parts, which in turn requires production runs of thousands of identical parts to recoup the high costs involved. In contrast to this is a more transient approach of evaluating multiple design options in digital design software, followed by direct manufacturing of one or more self-similar parts from the digital data. In the RoboFold process speeds are comparable to mass production, however every part can be different without any financial penalty; in essence it is a form of rapid prototyping similar to 3D printing.

Variation as Standard

At the 2012 Venice Architecture Biennale, Zaha Hadid Architects embraced the RoboFold technology with their dramatic sculpture *Arum* by creating an exciting new aesthetic using the system. The architects quickly realised the possibilities of a technology built from the ground up to embed the manufacturing characteristics in the design and control software. Output from generative and parametric design often seemingly offers 'unlimited' possibilities for architects, but all too easily defies the laws of physics and economics. Having developed a revolutionary new process for dealing with the physical constraints of folding metal with industrial robots, RoboFold must continue to advance research and development of the technology while maintaining an understanding that a balance between aesthetics, manufacturing and financial feasibility all contribute to its viability as an effective industrial process.

The Process

RoboFold has developed and released a suite of CAD software plug-ins to manage the design-to-production workflow. The software is based in the popular Rhino® and Grasshopper™ platform, and manages each stage of the workflow to enable a parametric link from start to finish. Design starts with quickly realising paper folding studies to ensure sheet material can be used from the outset. The development continues with a process of manual surface analysis that extracts the necessary data to simulate folding in KingKong, a Grasshopper plug-in using the Kangaroo physics engine, to create computational folding simulations as well as facade studies. The KingKong plug-in outputs the varied data in two forms: as flat patterns for cutting and as folding animations to drive the robot simulation. Cutting on the CNC router is facilitated by Unicorn, another Grasshopper plug-in, to generate the G-code CNC programming language. The Godzilla six-axis robot simulation plug-in is a powerful yet intuitive robot simulation environment in Rhino and Grasshopper where all the necessary checks for production feasibility occur. The final software stage sees Mechagodzilla take over and generate code for the robot on a remote Raspberry Pi®. Once the metal is cut and scored on the router, it is positioned below the robots that pick it up with vacuum end effectors, and the folding begins.

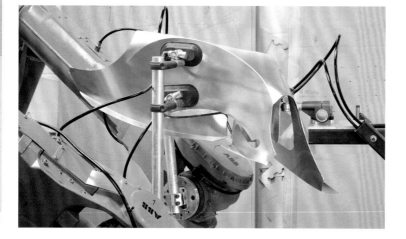

RoboFold, Sartorial Tectonics folded panel system, 2013
RoboFold developed a folded panel system based on textile folding patterns for Andrew Saunders of Rensselaer Polytechnic Institute (RPI) in New York. The project takes its inspiration from traditional pleating techniques, and specifically the box pleat. Folds in the cloth are translated directly to metal through material simulation and physical experimentation. A total of 11 panels were produced as a part of 1:1 facade mock-up, and shipped to the US to be assembled and exhibited at RPI.

Robots to the Core

RoboFold's client services are focused on developing production solutions that include software development, design consultancy, prototyping and licensing of its manufacturing technology for designers, artists, architects, and automotive, electronics and multinational industrials.

The company's experience in developing an end-to-end robotic fabrication workflow for the patented RoboFold process has now been encapsulated in its new Robots.IO (short for Robots: Input/Output) consultancy, which has an increased focus on high-value software and solutions for robotics professionals, again using Rhino and Grasshopper as a CAD platform.

Accessing robots has never been easier, and their possibilities are becoming more apparent. Robots.IO creates custom robotic solutions and continues to provide Godzilla for robot owners to do the same. Recent projects range from web-app controlled robots to 3D scan-driven CNC-milling plug-ins and robot installations in industrial settings. Though this area of business is separate from the RoboFold brand's cutting-edge metal-forming process – which is now also a customer of Robots.IO – both will continue to benefit from the company's continuously high level of investment in robotics research and development. ⌂

The company builds on the knowledge of how to work with the aesthetics and physics of sheet metal to achieve effective design solutions.

Zaha Hadid Architects, *Arum*, Venice Architecture Biennale, 2012
A 6-metre (20-foot) high sculpture comprising 488 load-bearing aluminium panels. Each panel is formed in the translation of a delicate human touch on paper into the decisive action of two six-axis robots. RoboFold manufactured the unique panels after a collaborative development phase with Zaha Hadid Architects and took charge of coordinating and executing the assembly on site.

Text © 2014 John Wiley & Sons Ltd. Images: p 68 © Gregory Epps/RoboFold Ltd; p 69 © Matthias Urschler

Andreas Froech

MACHINEOUS

Machineous is a specialised fabrication facility that has made the use of CNC equipment the focal point of its activity and exploration in the making of architecture. The process begins with an architect's CAD model and includes development of the design details, engineering, prefabrication, assembly and installation logic. Large industrial robots are the centrepieces on the factory floor and perform many of the initial steps to produce custom parts at many scales. The company has manufactured complex facade panels for Patrick Tighe Architecture and Kevin Daly Architects, experimental installation projects for Greg Lynn, Zaha Hadid, Hitoshi Abe and Tom Wiscombe, and furniture-scale products for Ammar Eloueini, Aranda\Lasch and Jeffrey Inaba.

Machineous, based in Los Angeles, was founded by Andreas Froech in 2008. Trained as an architect in Vienna, Austria and at Columbia University in New York, Froech developed his interest in the digitally supported fabrication of architecture when working with Greg Lynn and teaching at the University of California, Los Angeles (UCLA) School of Architecture between 1997 and 1999. Self-taught in the operation of CNC equipment, he was among the first to explore its potential for architectural fabrication. Prior to founding Machineous, he was Director of Material Development at Panelite where he was responsible for the development of several of the company's patented and award-winning composite honeycomb panels and systems, as well as for specialised material research contracted by OMA and Prada.

Robots are the company's equipment of choice due to their outstanding strength, speed, precision and reliability. They are used to operate plasma cutters, spindle rotary cutters and large circular saws. A single operator and a single PC station can interact with any of the robots to produce a large variety of parts efficiently. Many parts are one of a kind but share a common setup with respect to the equipment. An intense front-end software process allows the operator to identify common process methods, developing cost and material savings to fit most construction budgets. Up to 20 staff execute processes including wood- and metal-working, industrial painting and project management.

The extensive evolution of computer software-driven design allows architects to produce a wide variety of design solutions and to develop highly complex forms. The x,y,z data required to generate CAD construction drawings and presentation renderings can also be used to drive robotic equipment. Machineous has developed many proprietary conversion postscripts to translate that same data into actual robotic arm movement for the production of architectural parts.

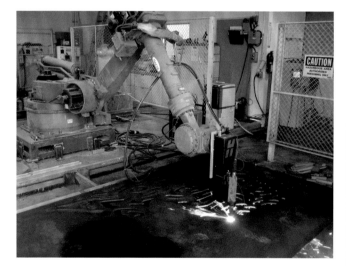

Machineous, Robotic fabrication facility, Gardena, California, 2013
Rapid metal cutting of a 9-millimetre (0.3-inch) steel plate using a CNC plasma cutting system. Production of elements for the Malibu Lagoon Canopy by Clark Stevens with Bunch Design.

Clark Stevens with Bunch Design, Malibu Lagoon Canopy, California State Parks, 2013
Final installation of the building's CNC-cut 9-millimetre (0.3-inch) steel plates.

Machineous operates with a deep understanding of the nature and behaviour of different materials, whether wood, plastic, metal or other, and of the behaviours, limitations and expressions of the robot. Fabrication processes and project-specific solutions are developed to exploit and embrace the interaction of specific material qualities with robotic technology.

In the past, such high-level technology and fabrication processes were mainly only available to projects with substantial budgets. However, recent reductions in equipment costs combined with Machineous's own efficient production and setup methods now enable affordable design solutions at any scale, from large metal facade screens with complex patterns, to freeform furniture and experimental installations.

Machineous is now enjoying rapid growth. It currently operates five large robotic stations integrated with a full service fabrication and finishing facility under one roof, and is geared towards further expansion to become one of the premier ornamental metal and specialist design fabricators in the US. ⌂

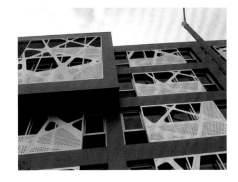

The x,y,z data required to generate CAD construction drawings and presentation renderings can also be used to drive robotic equipment.

Greg Lynn FORM, SITE, Santa Fe, New Mexico, 2012
above: Large moulded fibreglass facade components in free-form geometry.

Patrick Tighe Architecture, Sierra Bonita Affordable Housing, West Hollywood, California, 2013
top: The building's 125 CNC-cut 6-millimetre (0.2-inch) thick aluminium facade and balcony screen panels use 36 unique patterns.

Text © 2014 John Wiley & Sons Ltd. Images © Machineous LLC

Tobias Bonwetsch and Ralph Bärtschi

ROB TECHNOLOGIES

ROB Technologies provides software solutions that enable highly flexible digital fabrication processes for the efficient production of small batches of building components as well as individual non-standard parts. Founded in 2010, the firm has a staff of three who combine expertise in architecture and building knowledge, robotics and software development to offer software environments that drive the robots of different customers from industry as well as research institutions. Based in Zurich, it builds upon the knowledge and experience of its founders, Tobias Bonwetsch and Ralph Bärtschi, who for more than six years undertook pioneering research on robotic processes in architecture with Fabio Gramazio and Matthias Kohler at ETH Zurich, where together they hold the Chair of Architecture and Digital Fabrication.

At ROB Technologies, the industrial robot is understood not only as a means of achieving greater efficiency in production processes, but also, due to its inherent flexibility, as a tool that holds the potential to enhance a greater degree of freedom in design and construction. This potential combined with the ability to adapt to diverse production jobs and to realise a multitude of different fabrication processes, should make industrial robots the tool of choice within the building industry. However, in relation to the size of the market, the number of industrial robots actually performing building tasks is negligible.[1]

Generally, the arguments for highly flexible automated fabrication processes are indisputable: on the one hand sophisticated digital design tools are available, while on the other the dexterity of industrial robots enables the performance of arbitrary fabrication tasks. Unfortunately, though, in reality a gap still exists between the conceiving and planning of a design and its execution by (just in theory) highly flexible industrial robots. The problem seems to be located in the actual activation and utilisation of this flexibility. Robot manufacturers only provide proprietary old-style robotic programming languages that have a limited level of abstraction.[2] Industrial robots need to become easier to control and more intelligent in perceiving their environment. As the programmability of industrial robots is pivotal, ROB Technologies concentrates on the development and provision of software solutions.

To make the control of robotic systems more flexible and easier to use, thereby enabling companies to exploit these capabilities in their manufacturing tasks, ROB Technologies locates the starting as the design phase. The company provides fabrication-specific design tools that are combined with flexible control of robotic systems. Programming the robot and adjusting the system to modified outputs and processes can therefore be performed by non-experts in a fast and efficient way. By primarily targeting the construction industry, its core technology is also of high interest to other manufacturing industries that require automation but are reluctant to invest in traditional robotic solutions due to their complexity and costs.

Industrial robots need to become easier to control and more intelligent in perceiving their environment.

Gramazio & Kohler, Brick facade of Keller AG Headquarters, Ofenhalle, Pfungen, Zurich, 2012
The facade, part of the modification of a former production hall into the new headquarters of brick manufacturer Keller, was realised by applying the robotic assembly process developed by ROB Technologies.

One of ROB Technologies' exemplary products is BrickDesign, a comprehensive approach to the design and robotic fabrication of brick facades.[3] The software incorporates parameters of the robotic fabrication process already in the design phase, exploiting the capabilities of the robot to effortlessly position each individual brick differently. The data produced is directly used to control the fabrication process without the need for additional process programming of the robotic system. Efficient and highly flexible automated production of non-standard brick facade elements is thus realised.[4] Aside from fully realised prototypical facade projects like the Ofenhalle in Pfungen by architects Gramazio & Kohler (2012), the first large-scale commercial projects based on the BrickDesign software, in 2014, will be the Le Stelle di Locarno residential building built in Ticino, Switzerland, by Buzzi studio d'architettura, a further residential building in London, and sports facilities in Manchester, in 2014.

ROB Technologies' CAD-based URStudio software environment for off- and online programming of universal robots offers bi-directional communication. The goal is to ease complex task programming of the robot through a unique combination of teaching and manipulating virtual geometries, without the need to descend into programming machine code.

The aim of ROB Technologies is to foster the application of industrial robots in architecture by helping to activate their intrinsic potential, so that robots can become a truly powerful tool for innovation in the future of building and manufacturing. ᴀᴅ

Notes
1. The estimated number of industrial robots in the building industry is less than 50 robots per 10,000 employees, with most of them not performing actual building tasks, but handling jobs like palletising. See www.ifr.org/.
2. See A Hoffmann, A Angerer, F Ortmeier, M Vistein, and W Reif, 'Hiding Real-Time: A New Approach for the Software Development of Industrial Robots', *Proceedings of the 2009 IEEE/RSJ International Conference on Intelligent Robots and Systems (IROS)*, St Louis, Missouri, 2009, pp 2108–13.
3. The robotically assembled facade system was a finalist in the euRobotics Technology Transfer Award 2013. Its enabling software, BrickDesign, is available at www.food4rhino.com/project/brickdesign.
4. See Tobias Bonwetsch, Ralph Bärtschi and Matthias Helmreich, 'BrickDesign: A Software for Planning Robotically Controlled Non-Standard Brick Assemblies', in Sigrid Brell-Cokcan and Johannes Braumann (eds), *Rob|Arch 2012: Robotic Fabrication in Architecture, Art and Industrial Design*, Springer (Vienna), 2013, pp 102–9.

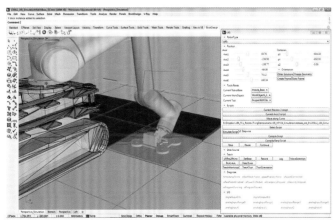

ROB Technologies, URStudio programming environment, 2013
Screenshot of URStudio controlling a tiling process. Ongoing research collaboration with the Future Cities Laboratory in Singapore.

Text © 2014 John Wiley & Sons Ltd. Images: p 72 © Gramazio & Kohler; p 73 © ROB Technologies AG

Ryan Luke Johns

GREYSHED

GREYSHED is just that: a 4 x 8 metre (14 x 25 foot) breeze block garage. From behind a rolling door in a row of postwar homes in suburban Princeton, New Jersey, a dull machine hum emanates. Inside, an assemblage of components salvaged from local scrapyards work to convert the common domestic power supply to the three-phase, 480-volt demands of a two-and-a-half-tonne industrial robot. 'Abraham' is a late-1990s welding robot that was stripped from a Ford factory line and acquired for less than $10,000 from a second-hand robot dealer found on eBay.

While the complications of transporting, recommissioning and operating such a behemoth are far from trivial, the unfamiliarity and scale of the machine belie its relative simplicity. GREYSHED aims to diminish the mystique surrounding this technology by expanding the territory of architectural robotics from the advanced institution into the unremarkable and unsupported suburban landscape. Just as the domestic appropriation of military technologies in the 1950s turned the marvellous to the mundane,[1] GREYSHED takes the robot off the pedestal and puts it in the carport.

The robot is not the future: it is already here. Under that assumption, GREYSHED explores complex design experiments with commonplace components. Recent projects turn used smartphones into augmented-reality headsets,[2] appropriate game console hardware to create gestural design platforms, and enable mediated tool-path manipulation with touchscreen tablets. Likewise, the robot's end effectors are hacked together from cheap, second-hand power tools. Precise fabrication is executed with an electric chainsaw, handheld router or bandsaw, glue or resin extruders are powered by a cordless drill,[3] and the vacuum-gripper runs on refrigerator parts. While the low-budget nature of such independent research can be limiting, the associated flexibility is simultaneously liberating. Essentially, GREYSHED is a decision to exchange institutional red tape for duct tape.

Just as the domestic appropriation of military technologies in the 1950s turned the marvellous to the mundane, GREYSHED takes the robot off the pedestal and puts it in the carport.

GREYSHED research lab and fabrication shop, Princeton, New Jersey, 2013
Founded in 2011 by Ryan Luke Johns and Nicholas Foley, GREYSHED is a garage-based collaborative focused on architectural robotics and design/fabrication workflows.

The balance between limited resources and resourcefulness characterises not only the means by which GREYSHED approaches its work, but the work itself. It is not the lack of constraints, but the variation of traditional constraints that enables a novel approach to design. The Robotic Poché project, for example, explores 'slow fabrication' procedures while simultaneously dealing with the practical problems of a poorly insulated, garage-based studio.[4] In order to decrease heat loss during winter research while reducing the noise transmission associated with fabrication work, the project engages material assemblies that provide both thermal and acoustic isolation – decoupled surfaces with foam infill. Here, the robot is used as a reconfigurable formwork for laying a complex configuration of ceramic tiles over the imprecise, pre-existing ceiling structure. By holding each tile in place while the void between the tile and the ceiling is filled with expanding polyurethane foam (and remaining in place until the foam cures), the finely tuned and inhumanly patient manipulator makes traditionally crude materials and processes viable tools for digital fabrication.

Operating somewhere between research and practice, GREYSHED fluctuates as it must from laboratory to design studio, consultancy and fabrication shop. Founded in 2011, it follows the entrepreneurial spirit of garage innovators by balancing research, play, production and collaboration. Propelled by polarity, it explores the space dividing the traditional architectural dichotomies of design/construction, digital/analogue, stochastic/deterministic, man/machine, simulation/execution and amateur/professional.

Through the simultaneous occupation of multiple phases of the design-production spectrum, GREYSHED seeks to create not only 'highly informed' architecture,[5] but highly informed architects. Operating at a localised scale of 'byte to robot' rather than 'file to factory', the design–prototype–production sequence is compressed into a feedback loop that empowers the designer to pre-empt problems generally faced by engineers and contractors long after the initial design impetus has passed. By fostering concurrent computation, construction, craftsmanship and design, GREYSHED works to advance digital fabrication while revitalising the role of the human designer. 𝔻

Notes
1. Beatriz Colomina, *Domesticity at War*, MIT Press (Cambridge, MA), 2007.
2. Ryan Johns, 'Augmented Reality and the Fabrication of Gestural Form', in Sigrid Brell-Cokcan and Johannes Braumann (eds), *Rob|Arch 2012: Robotic Fabrication in Architecture, Art and Industrial Design*, Springer (Vienna), 2013, pp 248–55.
3. Buoyant Extrusion: www.gshed.com/portfolio/buoyant-extrusion.
4. Nicholas Foley and Ryan Johns, 'Irregular Substrate Tiling: The Robotic Poché', in Brell-Cokcan and Braumann, op cit, pp 222–9.
5. Tobias Bonwetsch, Fabio Gramazio and Matthias Kohler, 'The Informed Wall: Applying Additive Digital Fabrication Techniques on Architecture', *Proceedings of the 25th Annual Conference of the Association for Computer-Aided Design in Architecture*, 2006, pp 489–95.

GREYSHED, Mixed Reality Modelling, 2013
By combining robotic manipulation with projective augmented reality, gesture tracking and computer simulation, the design and fabrication processes operate in a material- and designer-informed feedback loop.

GREYSHED, Irregular Substrate Tiling: The Robotic Poché, 2012
The incredibly patient robot is used as a reconfigurable formwork: holding each tile in place for almost an hour while the polyurethane bonding agent sets.

Text © 2014 John Wiley & Sons Ltd. Images: p 74 © GREYSHED, photo by Nicholas Foley; p 75 © GREYSHED, photos by Ryan Luke Johns

COMPUTATION
or
REVOLUTION

THE CONNECTION BETWEEN TECHNOLOGY AND SOCIETY TODAY REMAINS AS COMPELLING AS IN LE CORBUSIER'S TIME, WITH VIEWS FROM SOCIAL COMMENTATORS, SCIENTISTS AND ECONOMISTS HABITUALLY POLARISING AROUND THE ADVERSE OR POSITIVE IMPACT OF TECHNOLOGICAL CHANGE. TRACING THE ROOTS OF COMPUTATION AND ROBOTICS BACK TO THE SECOND WORLD WAR, **PHILIPPE MOREL** OF EZCT ARCHITECTURE & DESIGN RESEARCH REDEFINES THE POSITION OF ARTIFICIAL INTELLIGENCE, ROBOTS AND COMPUTATION IN ARCHITECTURE, HIGHLIGHTING THE POTENTIAL OF COMPUTERS TO PERFORM PRECISE CALCULATIONS – AND OUTPERFORM HUMAN INTELLIGENCE IN ALMOST EVERY WAY.

Philippe Morel

We must remember that another and a higher science … has given to us in its own condensed language, expressions, which are to the past as history, to the future as prophecy ….
It is the science of *calculation* – which becomes continually more necessary at each step of our progress, and which must ultimately govern the whole of the applications of science to the arts of life.

—Charles Babbage, 1832[1]

ARCHITECTURE OR REVOLUTION
… The machinery of Society, profoundly out of gear, oscillates between an amelioration, of historical importance, and a catastrophe.

—Le Corbusier, 1923[2]

In Le Corbusier's key work entitled *Vers une Architecture* [*Toward an Architecture*], published in 1923, he maintained that the world's transformation by machines would lead either to 'an amelioration, of historical importance', or to 'catastrophe'. Today, it is tempting to use these same terms when referring to the impact of the information technology revolution and of linking computation and machines in what we consider here, in general terms and without distinction, as 'robotics' or 'mechatronics' – the latter term referring to a fusion of mechanical engineering, electronic engineering and computer engineering. In fact, Le Corbusier's opinions, and more generally the French architect and theorist's studies on the connection between technology and society, are still so compelling that not a week goes by without a scientist, futurist or economist publishing on the social impact of technological change, as demonstrated by recent articles and works by Erik Brynjolfsson, Kevin Kelly and Paul Krugman.[3] All three seek, amongst other things, to revive certain Marxist theories on technology – including Krugman, whom no one would consider leftist.

However, although the debate on the role of technology and, more specifically, robotics in the multiple economic and social transformations taking place today is very topical, due mainly to exponential growth in the quantity of products available and a drop in production costs, it remains blurred. Firstly, it was recently subjected to a short-term productivity analysis which implicitly identified the IT revolution with human computer users rather than

From metaphorical physics to effectiveness
bottom: A juxtaposition of modernist belief in the power of scientific metaphors, expressed as a diagram (top), and a 'Space-time diagram of relativistic arbitrage transaction' (bottom). The latter, from AD Wissner-Gross and CE Freer's 'Relativistic Statistical Arbitrage' (*Physical Review E*, 82, 056104, November 2010, p 2), shows how '[s]ecurity price updates from spacelike separated trading centers (squares) arrive with different light propagation delays at an intermediate node (circle) at time t=0, whereupon the node may issue a pair of "buy" and "sell" orders back to the centers'.

Worldwide map of high-frequency financial trading
below: The map, from AD Wissner-Gross and CE Freer's 'Relativistic Statistical Arbitrage' (*Physical Review E*, 82, 056104, November 2010, p 5), shows optimal intermediate trading node locations (small dots) for all pairs of 52 major securities exchanges (large dots). Wissner-Gross and Freer explain: 'Recent advances in high-frequency financial trading have made light propagation delays between geographically separated exchanges relevant. Here we show that there exist optimal locations from which to coordinate the statistical arbitrage of pairs of spacelike separated securities, and calculate a representative map of such locations on Earth. Furthermore, trading local securities along chains of such intermediate locations results in a novel econophysical effect, in which the relativistic propagation of tradable information is effectively slowed or stopped by arbitrage' (ibid, p 1).

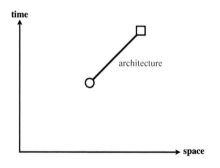

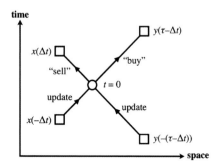

with the machines themselves. This obscured the defining feature of a technology which has since opened up the path to full automation in production and to algorithmic capitalism in finance. To a large extent, this is a recent form of capitalism and its development, conversely, tends to validate the assertion by Paul Krugman that 'Productivity isn't everything, but in the long run it is almost everything.'[4]

Following on from this, the debate on the social impact of computation came up against a remarkable lack of knowledge concerning the already distant origins of information technology and consequently the concepts which relate to it. This lack of knowledge can be attributed to disdain for the technology or to the difficulty in accessing an aspect of modern culture based largely on science. Although the conceptual basis was laid prior to the 20th century, the operational aspects of information technology and electronics were developed in an era much closer to the publication of *Toward an Architecture* than to the modern day. This was an era associated with the Second World War, whose importance can scarcely be overestimated: *anni mirabiles* 1947 and 1948 alone saw the invention of the transistor and the publication of the founding theories of contemporary science and technology by John von Neumann, Claude Shannon and Norbert Wiener.[5] As is illustrated by the *Fortune* magazine covers shown here, the technological revolution following the

 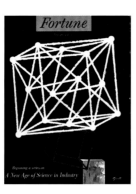

Fortune magazine covers, June 1946 (left) and January 1953 (right)
left: Published by Time Inc, the titles of these issues of *Fortune* magazine – 'Fundamental Science' and 'A New Age of Science in Industry' – illustrate the importance of science in industry.

The Origin of our World: key publications of 1948
below: During the years 1947 and 1948, the foundational books of computational sciences were published, as well as probably the most important book written by an architectural historian in the 20th century (*Mechanization Takes Command*, by Sigfried Giedion), even if this book did mention mechanisation precisely at the time when we were entering the age of computation. From top left to bottom right: *Theory and Techniques for Design of Electronic Digital Computers*, Vol III, by John von Neumann, 1948; *A Mathematical Theory of Communication* by Claude Shannon, 1948; *Cybernetics or Control and Communication in the Animal and the Machine* by Norbert Wiener, 1948; and *Mechanization Takes Command* by Sigfried Giedion, 1948.

Second World War has resulted from a very deep intricacy between fundamental science (including mathematics and quantum physics) and industry. Our robotic age is also characterised by such an association, probably at an even deeper level.

Lastly, agreement, even among bona fide experts, on the future development of a society, indeed a whole civilisation, is not necessarily a given, as shown by the differences of opinion between Arthur W Burks and Douglas Hartree on the future of information technology. The former predicted, after the creation of the Electronic Numerical Integrator and Computer (ENIAC) – the first general-purpose computer, developed by the United States Army's Ballistic Research Laboratory to calculate artillery firing tables in 1946 – that 'a new era, an era of electronic calculation' was starting;[6] while the latter, who took a more reserved approach, foresaw the commercial failure of computers, saying that no country could ever train the number of programmers needed to sell computers.[7] These two attitudes still featured in the debates that shook the 1960s and 1970s and which, added to those addressed further on in this piece that examine the impact of technology from the work point of view, mirror the current discussions on robotics.

The Market of Robotics

The robotics market is currently growing at a phenomenal rate, driven by progress in all areas of industry from electronics and precision machining to materials science and systems engineering. This impacts on the performance and cost of robots, paving the way for a whole range of new applications. For example, the cost of industrial robots, adjusted according to quality, decreased from a factor of 100 to 20 between 1990 and 2004,[8] while sales of these same robots increased by a yearly average of 9 per cent between 2008 and 2012, with a record total of 166,000 robots sold in 2011.[9] As regards non-industrial robots, for instance unmanned aircraft and vehicle systems (UASs and UVSs) used in precision agriculture, telecommunications, transport, surveillance and monitoring, a recent study estimates at $27.6 million the loss per day to the United States of failure to integrate these into the National Airspace System (NAS),[10] which gives a sense of the issues involved. Although architecture

 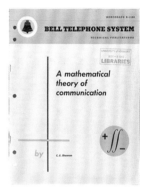

> THE ROBOTICS MARKET IS CURRENTLY GROWING AT A PHENOMENAL RATE, DRIVEN BY PROGRESS IN ALL AREAS OF INDUSTRY FROM ELECTRONICS AND PRECISION MACHINING TO MATERIALS SCIENCE AND SYSTEMS ENGINEERING.

An automated load haul dump truck operating underground
bottom: Today's GPS-based autonomous haulage systems, controlled from offices distant by thousands of kilometres, confirm what Philippe Morel wrote 11 years ago about the importance of 'information in its relationship with nature, its scientific, economic and political relationship' (in 'Le paysage biocapitaliste', *Revue Minotaure* (Paris) 2003).

is one of these issues and is the main focus here, assessing robotics in a wider context provides insights that are not visible in the early stages of development of architecture-related robotics.

Firstly, the actual definition of the word 'robot' and the different types of robot should be clarified. Until 2011, ISO standard 8373, issued by the International Standard Organisation (ISO), that defines terms used in relation to robots and robotic devices, defined an industrial robot as 'an automatically controlled, reprogrammable, multipurpose manipulator programmable in three or more axes, which may be either fixed in place or mobile for use in industrial automation applications'. Service robots, conversely, had no strict or formally accepted definition. This definition of industrial robots – still very similar to that of the Robot Institute of America used until 1993, which held that 'a robot is a reprogrammable, multifunctional manipulator designed to move materials, parts, tools, or specialized devices through variable programmed motions in the performance of a variety of tasks' – only partially reflected the development of robotics. Leaving aside service robots and non-standard robots, the definition continued to be influenced by the industrial context of a robotics technology still in its infancy. This era was symbolised by the term 'Article Transfer', taken from the title of a patent filed on 10 December 1954 by George Devol named *Programmed Article Transfer*. (Granted by the United States Patent and Trademark Office on 13 June 1961, the patent gave rise to the first industrial robot, Unimate, manufactured by Devol's company Unimation and put into use in the latter year.) The term 'Programmed' confirmed the stronger link with computers. While the key notions of 'automatic' and 'reprogrammable' were maintained in the standard definition of robots following on from this first patent, the other terms excluding service robots gradually became increasingly problematic. These robots, which do not have the kinematic or generic design of industrial robots, are nevertheless more popular today. This was one of the factors that led to the update in 2012 of ISO 8373, whose

general definition of a robot is an: 'actuated mechanism programmable in two or more axes … with a degree of autonomy …, moving within its environment, to perform intended tasks. Note 1 to entry: A robot includes the control system … and interface of the control system. Note 2 to entry: The classification of robot into industrial robot … or service robot … is done according to its intended application.'

These remarks could be considered linguistic subtleties if they did not show how great the trend now is to class any object, be it just slightly mobile or intelligent, as a robot, leading to a Suprematist-type world without objects. If the universal Turing machine was the reference point used to evaluate the computational capacity of different computers and therefore their

> WHILE THE KEY NOTIONS OF 'AUTOMATIC' AND 'REPROGRAMMABLE' WERE MAINTAINED IN THE STANDARD DEFINITION OF ROBOTS FOLLOWING ON FROM THIS FIRST PATENT, THE OTHER TERMS EXCLUDING SERVICE ROBOTS GRADUALLY BECAME INCREASINGLY PROBLEMATIC.

1998-generation Navlab robots, Houston, Texas
top: Their name a contraction of 'National Autonomous Vehicle Laboratory', the Navlab 9 and 10 driverless buses were developed in collaboration with the Metropolitan Transit Authority of Harris County, Houston.

A job application from an ideal worker: the Unimate 2000B
opposite: Having secured the necessary patent in 1961, George Devol's company Unimation, Inc brought out the first industrial robot, naming it Unimate. This spoof job application was produced by Unimation in collaboration with IFS Publications Ltd, c 1970.

VITA

UNIMATE 2000B

Shelter Rock Lane
Danbury, CT 06810

Social Security No.: None
Age: 300 hours
Sex: None
Height: 5 ft.   Weight: 2800 lbs.
Life Expectancy: 40,000 working hrs.
                (20 Man-shift yrs.)

Position Desired: Die Cast Machine Operator

Salary Required: $4.00/hr.

Other Positions for Which Qualified:
Forging press, plastic molding, spot welding, arc welding, palletizing, machine loading, conveyor transfer, paint spraying, investment casting, heat treatment, etc.

Education: On the job training to journeyman skill level for all jobs listed above.

Languages: Record-playback, Fortran, assembly

Special Qualifications: Strong(100 lb. load), untiring 24 hours per day, learn fast, never forget except on command, no wage increase demanded, accurate to 0.05" throughout range of motion, equable, despite abuse.

History of Accidents or Serious Illness:
Suffered from Parkinson's Disease(since corrected), lost hand(since replaced), lost memory(restored by cassette), hemorrhaged(sutured and fluid replaced).

Physical Limitations: Deaf, dumb, blind, no tactile sense, one armed, immobile.

Notify in Emergency: Service Manager, Unimation Inc., (203) 744-1800

Dependents: Human employees of Unimation, Inc.

References: General Motors, Ford, Caterpillar, Bobcock Wilcox, Xerox, and 65 other major manufacturers.

Original patent of the first industrial robot, Programmed Article Transfer, by George Charles Devol, Jr, issued 13 June 1961. Devol applied for the patent on 10 December 1954, the document extending over a mere three pages.

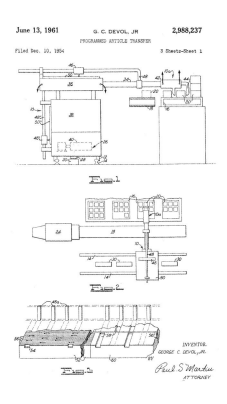

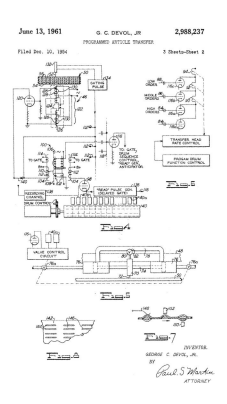

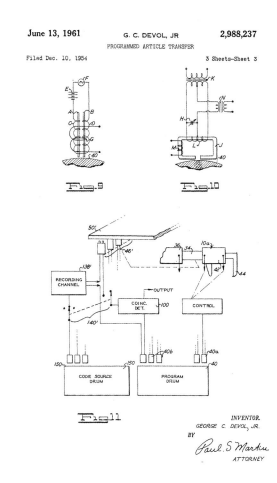

classes, the industrial robot, whose 6 degrees of freedom in linear movements and rotations (3+3) allow it to reach any possible point within its work envelope, represents a reference electromechanical machine, a mechatronic benchmark for the robotics era. Although the old distinction between industrial and service robots still exists to a certain extent, the boundary is, as mentioned, porous in the sense that it is the application that determines the classification. For example, Section 2.10 of ISO 8373 (2012) states: 'While articulated robots used in production lines are industrial robots, similar articulated robots used for serving food are service robots.' It is not difficult to transpose this kind of distinction to architecture, but in doing so, answering this simple question becomes a challenge: what do we mean when we talk about robotics in architecture and how will it or should it be distinctive?

Specificity of Robots or Specificities of Architecture?

Architecture-related robotics could potentially be split into 'architectural robotics' and 'robotic architecture'; but does this first branch of robotics actually exist? If logistics and transport have, for example, led to the development of new kinds of robots or, at the very least, to the classification of mobile machines (which until recently could only be called cars, tractors or planes) as robots, the same innovation process seems much slower in architecture. Indeed, no specific area of robotics has been adapted for use in architecture. To date, experiments in automating prefabrication carried out in Japan mainly in the 1970s and 1980s and, more recently, in Western so-called 'avant-garde'[11] architecture have mainly relied either on standard industrial robots or non-standard but existing robots, such as unmanned aerial vehicles (UAVs). In the first case the end result is still similar to what it would be without the use of automated processes, while in the second case, with more innovatory features, innovation is more tangible in architecture than in robotics itself.

Although 159,000 industrial robots were sold throughout the world in 2012 alone and the number of service robots in use has exceeded the 10 million mark, a very small proportion of them have found their way into architecture. If the construction materials and parts industry is highly, indeed for certain plants entirely, automated and if service robots are already being used on construction sites, an increase in productivity, within a fully liberal economic system, is surely the likely result. So which criteria distinguish between the advanced architectural research mentioned here and the simple opportunistic applications used by industry? First of all, steps should be taken to combine a revolution of architectural and constructive language with a revolution of social organisation which is now out of step with the general transformations taking place in science and technology. If possible, this should be done, as stated by Paul Krugman, 'before the robots and the robber barons turn our society into something unrecognizable'.[12] Secondly, there should be a clear awareness that robotics in architecture cannot be isolated from this same society – that, despite appearances and obviously beyond what it represents as a specific area of science, robotics constitutes neither a procedural problem, nor a discipline problem, nor even a work problem, but rather an aspect of the generic problem of our era: computation. Although, when returning to the source of cybernetics, robotics appears as a control problem and, when adding an additional level of abstraction, as a logical-mathematical problem, in both cases they require their own epistemology and historical understanding. Yet, more often than not, this is quite simply missing in architecture. While Charles Babbage's theorising on the unavoidable

Three stages of Suprematism: the geometric pop, the animal pop and the robotic pop
On the bottom we recognise the 100 *Suprematist Surface Forms* (1921–2) by Nikolai Suetin, on the left the Cow Wallpaper (1966) by Andy Warhol and below the Swarm of Nano Quadrotors at the GRASP Lab, University of Pennsylvania (2012).

replacement of work by machines in his 1832 book *On the Economy of Machinery and Manufactures* shows that he had learnt from the practices of division of high-skill labour advocated by French mathematician and engineer Gaspard de Prony, 36 years his senior, it seems little has been learnt since his time. This is in spite of the repeated warnings given by those closest to the scientific foundations of technical development.[13] As Kevin Kelly cautions: 'The rote tasks of any information-intensive job can be automated. It doesn't matter if you are a doctor, lawyer, architect, reporter, or even programmer: The robot takeover will be epic.'[14]

Does this mean that there is no specifically architectural problem linked to robotics? Of course it does not. All those who work with robots know, for example, that programming numerous non-repetitive trajectories in a reasonable time frame, however, is a problem. Developing software to reduce this work and use trajectories from standard CAD software, with automatic code generation if possible for several brands and types of robot, is another. Similarly, the incompatibility of construction materials and techniques from a pre-robotics era also presents its own difficulties, as does large-scale construction using robots most of which were not developed for this purpose. Nevertheless, without underestimating the work and intelligence needed to overcome these difficulties, they still appear to provide little indication about the development of robotic architecture, since they can be resolved in the short term. An example is the work towards simplifying communication between industrial robots and the types of CAD software most widely used in architecture that has been carried out at EZCT Architecture & Design Research since July 2010, when an ABB IRB 120 robot was acquired for the purpose. The consequent development of HAL, a Grasshopper™ plug-in and future standalone application for industrial robot programming and control,

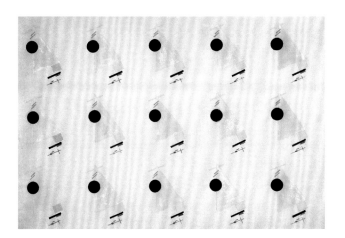

by EZCT intern Thibault Schwartz goes a long way to addressing the automatic code generation issue. The research conducted in the DFAB facility at ETH Zurich on adaptability to existing and non-calibrated materials has also been conclusive, as has the use of artificial intelligence and vision systems – now found in most service robots from the vacuum cleaner to the driverless car and including the drones used in precision agriculture – to manage dynamic tolerance problems. Scientific and technological discoveries have played a major role in the development of robotics, over and above the applications intended for them today. Although the pace of discovery in all areas is unprecedented in history, the following claim by Francis Bacon, who was certainly the first philosopher to have considered a theory of invention (with some scientific gaps), still holds true: 'If, for instance, before the discovery of cannon, one had described its effects in the following manner: There is a new invention by which walls and the greatest bulwarks can be shaken and overthrown from a considerable distance, men would have begun to contrive various means of multiplying the force of projectiles and machines by means of weights and wheels. … But it is improbable that any imagination or fancy would have hit upon a fiery blast, expending itself so suddenly and violently.'[15]

Artificial Intelligence and Computation as 'Communism of Genius'

The romantic associations between architecture and technology deflect industrial economy problems onto an aesthetic theory of widespread creativity. The theory holds that robots and machines are the Neo-Ruskinian gadgets of those types of architect who are incapable of accepting their own obsolescence, as convinced of their own infallibility as booksellers as they are of the need for a greater human touch than the algorithms of Google and Amazon. Of course, reality proves every minute that the opposite is true, and a mere glance at the situation architecture finds itself

in shows the extent of this misjudgement. Indeed, virtually all architecture can be described as a pastiche business, in which a few often basic outlines or elementary lines of scripts result in tens of thousands of buildings, just as the sperm of a few prime bulls results in millions of other bulls intended for the food chain.[16]

Saying that most political lessons of the emergence of computation have not been learnt is a euphemism in today's architectural practice and theory. Whether it is in terms of pure ability to perform calculations, as reflected in the word 'computer', or by means of artificial intelligence, which has a close but complex connection with calculation, computation – as an authentic product of the knowledge democracy or, as the Surrealists might have called it, the 'communism of genius' – 'overturns all our concepts of

> MOST OF WHAT WE CONSIDERED OR NOSTALGICALLY STILL CONSIDER AN ARCHITECTURAL ISSUE NO LONGER APPEARS AS SUCH WHEN EXAMINING THE LINK BETWEEN ARTIFICIAL INTELLIGENCE AND AUTOMATED PRODUCTION.

From the mechanics of Le Corbusier to those of the Robotic Age
The left image, a low-pressure air turbine produced by the Rateau factory, was published by Le Corbusier in *Towards an Architecture* (1923). The top image represents a Mecanum wheel, invented in 1973 by Bengt Ilon. Its main characteristic is that it can move a vehicle in any direction, therefore making it highly adapted to combinations of rotational movements. As important as the invention of the Cardan joint by Girolamo Cardano c 1545, it appears to us as the wheel of the robotic or computational age.

'Is Surrealism the Communism of Genius?', 1924
opposite: This Surrealist tract was published by the Bureau de recherche surréaliste in December 1924.

> **LE SURREALISME EST-IL LE COMMUNISME DU GÉNIE ?**
>
> Bureau de recherches surréalistes
> 15, rue de Grenelle, de 4 h. ½ à 6 h. ½

[architectural] culture'.[17] Most of what we considered or nostalgically still consider an architectural issue no longer appears as such when examining the link between artificial intelligence and automated production. For this, it is simply a question of changing perspective and observing a few recent phenomena within clear epistemological reach. Examples include the automated drafting of press articles from Reuters and Associated Press wires, articles which could not be distinguished from pieces written by humans and which recall Turing's work on the Imitation Game, or the recent developments in automated verification of mathematical theorems which even a 'traditional' mathematician like Vladimir Voevodsky considers fundamental. This progress, which led Voevodsky to claim that 'soon mathematicians won't consider a theorem proven until a computer has verified it',[18] validates the long-standing opinion, which has been taken up by epistemologist Franck Varenne in an engineering context, that simulation models are better than real models.[19] To sum up, in architecture, most of us still view the relationship between architecture and machines, more specifically robots, in a traditional framework at the very moment when it is disappearing.

Rather than wondering how robotics can play a part in an architectural discipline that is still romantically defined by a focus on ornament or architectural composition – which any well-programmed smartphone can determine in a fraction of a second without distinction as to the authorship – it has become clear that we must address artificial intelligence and accept that robots are first and foremost computers. Therefore, due to the capacity robots have to perform calculations, we can and must consider their potential in much more detail than we are currently.

Identifying the true nature of technologies has been a trademark of the best architects/theorists of the 20th century, for example Le Corbusier, Buckminster Fuller, and indeed Marshall McLuhan, who, although not directly concerned with the production of objects, wrote in 1969 – well before architects began to express a similar opinion – that 'once electronically controlled production has been perfected, it will be practically just as easy and affordable to produce a million different objects as to create a million copies of the same object'.[20] So, due to an exponential drop in the cost of electronics and thus technologies, a whole range of new approaches to robotics in architecture will start to appear. The opportunities presented by UAVs for transporting heavy loads are already among them, even though, as we have already said, most applications remain similar to what robotics offered six decades ago, namely Programmed Article Transfer.

In fact, real developments require a conceptual leap, for instance when the modern car was assembled from separate elements – horses and carriage – into a single object or when the flight of birds stopped being used as inspiration for aeroplanes and was replaced by a unique application of physical principles, which Le Corbusier delighted in reiterating. Automation and the constantly developing trend towards autonomous objects now give full meaning to his term 'machine for living', since the home is no longer an object built by robots but is itself potentially part of the 'robots for living' class, a class which was partly anticipated during the 1960s. The process started by Le Corbusier and the Russian productionists would therefore be completed, whereas at the same time the process started by the Futurists is constantly re-starting, each minute confirming the extent to which futurism is the unspoken and unwritten doctrine of our civilisation. A doctrine which the 'science of calculation' or to use the current term 'computer simulation' has implemented through 'its own condensed language, expressions, which are to the past as history, to the future as prophecy'.

A Temporary Conclusion on Computational Literacy and Politics

The time will come when architectural design and construction will only be taken seriously if entirely automated and checked by a computer. The prospect of this moment will be daunting to certain people, even though only a tiny percentage of buildings can actually be called architecture and global developments are governed not by this percentage but by the general majority. So architectural robotics should mainly target this group, not only because it can, but because it has no choice: as mentioned above, since 'the rote tasks of any information-intensive job can be automated',[21] they will be. In order to properly meet needs, the whole discipline should already be reviewing its practices and its bases, including the method of teaching, where the deficit in scientific knowledge is difficult to understand. If it takes a real author to write *The Brothers Karamazov* or a real architect to plan the Villa Savoye

Cover of the *Automated Trader Quarterly*, Fall 2011
This issue of the magazine was dedicated to the 'Rise of the Robot Market: Machines on the Alpha Horizon'.

> WHILE NO ONE NOWADAYS COULD IMAGINE LOOKING FOR A DOCUMENT WITHOUT THE HELP OF A SEARCH ENGINE, MANY PEOPLE STILL THINK THAT PROBLEMS WHICH ARE VASTLY MORE COMPLEX, FOR EXAMPLE IN POLITICS, ARCHITECTURE AND URBAN PLANNING, CAN BE RESOLVED 'TRADITIONALLY', IN OTHER WORDS BY HUMANS.

and to invent the concepts of which it is an ultimate proof, what is there to say about mere 'newspaper article architectures' where each one is desperately seeking an inaccessible originality of style? In an era of complexity, should we not be encouraging the only intelligence capable of tackling this complexity, namely artificial intelligence? While no one nowadays could imagine looking for a document without the help of a search engine, many people still think that problems which are vastly more complex, for example in politics, architecture and urban planning, can be resolved 'traditionally', in other words by humans. It is a disastrous belief, which the Russian theorists denounced almost a century ago when they asserted that 'talent is no longer random but artificially cultivated'; the 'reform of science and education' would therefore need to be 'viewed and implemented from a Constructivist angle',[22] which, today, translates as foregoing the simple algorithmic aesthetics approach for the assimilation of these principles. If architects had the opportunity to use both automated construction and design, they would place themselves much higher up the scale of production.[23] They would also adapt their current superficial use of technology, for example that of electronics which is becoming more financially accessible just as architecture is moving at a forced march in the opposite direction (which incidentally invalidates cynical low-tech approaches) …

Based on these various points, it would appear that there are three challenges facing automation in architecture and indeed automation in general. The first challenge is to ensure that digital and computational literacy is properly integrated into teaching at architecture schools, as of first year and then at various subsequent stages of learning. The need to acquire and transfer this new type of knowledge in the education system led me, among

other reasons, over 10 years ago to stop using architecture software in favour of scientific computing software, i.e. Mathematica®. Following on from the key work carried out by DFAB at ETH Zurich, the second challenge is to test and develop new types of robot which would not only allow such a work to be replicated but which would move architecture forward; for although it is an age-old discipline, architecture is needed now more than ever. As a result, together with the COPRIN team at INRIA Sophia Antipolis and ENSAM Paris, we are currently testing a Stewart platform-type parallel robot. The robot, which is portable and cost-effective, can in theory cover an area of several thousands of square metres while bearing a load of more than 500 kilograms (1,100 pounds). Its design involves closed-loop kinematic chains that ensure good rigidity and high repeatability; however, it is difficult to model the behaviour of this kind of parallel robot since the computation of a configuration requires the solving of a set of non-linear equations which theoretically may have up to 40 solutions, this computational complexity involving a couple of non-trivial problems for low-cost implementations. The third, much less specific challenge implies a general understanding of the nature of robotics as computation applied to objects.

If the refusal to adopt this wider application of computation in architecture sometimes appears to be a stumbling block, more often than not this is the result of a pragmatic or more precisely opportunistic use of the latest technological advances in an unsuited social context. This is a 'prehistoric era with oversupply',[24] where false political and social foundations are upheld by capitalism itself or by subjects whose acceptance of traditional politics, despite the accumulated evidence of failure, is a mystery. In most areas, artificial intelligence and automation are still only tentatively promoted and indeed kept separate from each other. The illusory need for certain intellectual skills is artificially maintained by entrusting design to 'human robots' that are underpaid but still profitable. The already effective automation of physical tasks combined with the intrinsic capacity of computers shows that architecture is lagging behind. In the same way that traditional trading floors represent the folklore of finance rather than finance itself when compared with robot-traders, architecture as we regrettably continue to conceive of it will unfortunately also become folklore. That being the case, we would have everything to gain – or at least not much to lose – in architecture as well as in all economy-related questions, by heeding Charles Babbage, and others more recently, and entrusting operations to calculation (Babbage's use of which term corresponds to today's 'computation'). Indeed, by solving the arithmetic side of the economic calculation problem famously considered by Ludwig von Mises as the cause of the failure of socialism, the computational power of Google servers and personal supercomputers proves to us at every second that artificial intelligence-based subsitutes are capable of managing a more rational and egalitarian society.[25] ⌂

Notes
1. C Babbage, *On the Economy of Machinery and Manufactures*, Charles Knight, Pall Mall East (London), 1832, p 316.
2. Le Corbusier, *Vers une Architecture* (Towards a New Architecture), Les Editions G Crès et Cie (Paris), 1923, p 227 of second edition with new preface.
3. See E Brynjolfsson and A McAfee, *Race Against the Machine: How the Digital Revolution is Accelerating Innovation, Driving Productivity, and Irreversibly Transforming Employment and the Economy*, Digital Frontier Press (Lexington, MA), 2012; K Kelly, 'Better Than Human: Why Robots Will – And Must – Take Our Jobs', *Wired*, 12 December 2012; and P Krugman, 'Robots and Robber Barons', *New York Times*, 9 December 2012.
4. P Krugman, *The Age of Diminished Expectations: US Economic Policy in the 1980s*, MIT Press (Cambridge, MA), 1990, p 9.
5. *Theory and Techniques for Design of Electronic Digital Computers*, a series of 48 lectures held at the University of Pennsylvania from 8 July to 31 August 1946 (including Lecture 40 by John von Neumann), transcripts published by the University of Pennsylvania (Philadelphia). The image published here is of Vol III, of June 1948; C Shannon, 'A Mathematical Theory of Communication', *Bell System Technical Journal*, 27(3), 1948, pp 379–423; and N Wiener, *Cybernetics, or Control and Communication in the Animal and the Machine*, MIT Press (Cambridge, MA), 1948.
6. AW Burks, 'Electronic Computing Circuits of the ENIAC', *Proceedings of the IRE*, 35, 1947, pp 756–7, 767.
7. 'I went to see Professor Douglas Hartree, who had built the first differential analysers in England and had more experience in using these very specialized computers than anyone else. He told me that, in his opinion, all the calculations that would ever be needed in this country could be done on the three digital computers which were then being built …. He added that the machines were exceedingly difficult to use, and could not be trusted to anyone who was not a professional mathematician …. It is amazing how completely wrong a great man can be. The computer business has since become one of the biggest in the world.' Lord Bowden, 'The Language of Computers', *American Scientist* 58, 1970, p 43.
8. 'A Trend in Robot Prices', from *Industrial Robotics*, a study by Professor Alessandro De Luca, Dipartimento di Informatica e Sistemistica 'Antonio Ruberti', Università di Roma 'La Sapienza', 2012.
9. Press release issued by the statistical department of the International Federation of Robotics, July 2013.
10. *The Economic Impact of Unmanned Aircraft Systems Integration in the United States*, a study by the Association for Unmanned Vehicle Systems International, March 2013.
11. This term is used by default and only for its symbolism. The current context of permanent revolution, at least in science and technology, cannot by definition have an avant-garde component.
12. P Krugman, op cit, p 9.
13. For example Norbert Wiener when he warned, without success, US Union representatives about the difficulty for workers to compete with machines. Many of Wiener's concerns are expressed in his two most famous books: *Cybernetics*, 1948, and *The Human Use of Human Beings*, 1950.
14. K Kelly, op cit: www.wired.com/gadgetlab/2012/12/ff-robots-will-take-our-jobs/all/.
15. F Bacon, *Novum Organum*, 1620, quoted in the edition by Joseph Devey (ed), PF Collier & Son (New York), 1902, p 85.
16. See P Morel, 'Living in the Ice Age', Master's thesis, available at ENSA Paris-Belleville Archive, 2002. And P Morel, 'Research on the Biocapitalist Landscape', the complete case study presented at the Archilab 2004 exhibition and partially available in *Archilab: The Naked City*, Editions HYX (Orléans), 2004, and *Verb Natures*, Actar (Barcelona), 2007. 'Le paysage biocapitaliste' (theoretical essay) was published in French in *Revue Minotaure* (Paris), 2003.
17. F Nietzsche, *Fragments posthumes sur l'éternel retour*, Ed Allia (Paris), 2003.
18. J Rehmeyer, 'Voevodsky's Mathematical Revolution', *Scientific American*, 1 October 2013.
19. F Varenne, 'What Does a Computer Simulation Prove? The Case of Plant Modeling at CIRAD (France)', in N Giambiasi and C Frydman (eds), *Simulation in Industry, Proceedings of the 13th European Simulation Symposium, Marseille, France, October 18–20, 2001*, SCS Europe BVBA (Ghent), 2001, pp 549–54.
20. M McLuhan, *Mutations 1990*, 1969. The wording used here is a translation from the French translation by François Chesneau, Mame (Tours), 1969, p 43.
21. K Kelly, op cit.
22. N Tarabukin, *From Easel to Machine* (Ot molberta k machine), Moscow, 1922–3.
23. See M Tafuri, 'Architecture and Utopia: Design and Capitalist Development', *Progetto e Utopia*, Laterza (Bari), 1973. Tafuri called for a similar ideological transformation in architecture, one that is difficult to instigate on a theoretical and practical level. Due to the complacency of architects, the Italian historian's theory has almost totally failed in practice.
24. In. 'Domination de la nature, idéologies et classes', *Internationale Situationniste* 8 (Paris), 1963, p 3.
25. On this debate and about the possibility of a computational socialism, see the work of Allin Cottrell and W Paul Cockshott.

Text © 2014 John Wiley & Sons Ltd. Images: p 78(tl) © Philippe Morel; p 78(bl&tr) © American Physical Society, from Alexander D Wissner-Gross, et al, 'Relativistic Statistical Arbitrage', *Physical Review* E 82, 056104 (2010); p 80(l) © MacDonald, Dettwiler and Associates, Inc; p 80(r) © The Robotics Institute, Carnegie Mellon University; pp 81, 84-5 courtesy Philippe Morel; p 83(l) © DACS 2014; p 83(tr) © 2014 The Andy Warhol Foundation for the Visual Arts, Inc/Artists Rights Society (ARS), New York and DACS, London; p 83(br) © KMel Robotics

Thomas Bock and Silke Langenberg

Cham
Building

Building Sites
Industrialisation and Automation of the Building Process

The introduction of robotics in construction is part of a much longer history of industrialisation and automation on the building site. **Thomas Bock** *of the Technical University of Munich and* **Silke Langenberg** *of the University of Applied Sciences, Munich, highlight how the Industrial Revolution and the development of a transport infrastructure in the 18th and 19th centuries in Europe first triggered the shift in the building trade from a largely localised industry into a national and mechanised one, leading to the highly advanced automated construction techniques that continue to be developed in Japan and other Asian countries to this day.*

The Industrial Revolution changed the building process, by then largely dependent on a local base of materials, skills, building knowledge and tradition, irrevocably. In the late 18th century and throughout the 19th century, new machinery, serial-produced elements and industrially fabricated materials started to appear on building sites, complementing long-term approved construction techniques. It was not, however, until the 20th century that there was a real attempt to adopt industrial manufacturing processes. By the 1920s and 1930s a few prototypical buildings had been realised, anticipating the so-called 'industrialised construction' processes that were rolled out at a larger scale during the second half of the century: from the serial prefabrication of building elements to the mass production of standardised housing estates and system buildings. The specialised robotic machinery and automated high-rise construction sites that were developed in Asia during the 1990s can be viewed in the context of a greater trajectory of mechanisation and industrialisation.

The beginning of the 21st century could again prove pivotal for the building process, with robotic fabrication having the potential to change the building site once more. On that account, it seems crucial to take a look at the development, influence and results of some historical precursors in order to understand that the implementation of robotics in architecture at a larger scale may not just require a first phase of experimental research and prototyping, but also a fundamental change in the early design stages as well as in the construction process that goes far beyond imitating existing building technologies.

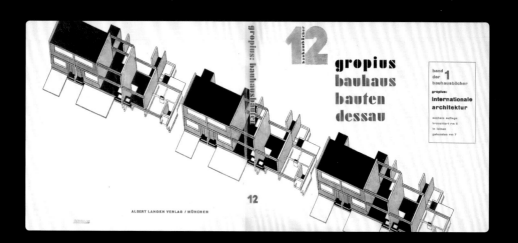

Joseph Paxton, The Crystal Palace, London, 1854
previous spread: The Crystal Palace was originally built in Hyde Park in London for the Great Exhibition of 1851. View from the south gallery of the Crystal Palace at Sydenham; from a photograph by HP Delamotte, from the *Illustrated London News*, 12 November 1853.

Walter Gropius, Cover of *Bauhaus Bauten Dessau*, 1930
Cover of *Bauhaus Bauten Dessau*, published by Gropius in Munich. Gropius used the Törten housing estate near Dessau to develop, test and promote the serial production of building elements with the intention of rationalising the building process.

The Establishment of Industrially Produced Elements and New Materials

The mechanisation that was a consequence of the Industrial Revolution has, by the expansion of the railway networks, thus increasing mobility and transportation of goods and materials, directly resulted in the growth and congestion of industrial towns, and the fast development of their building stock and infrastructure. It has also directly affected the building industry and construction process: industrially produced materials and prefabricated building elements such as cast-iron beams and columns, glass, factory-made bricks or artificial stone decorative elements were increasingly used during the 19th century, and their potential in the erection of large industrial, infrastructural and representational buildings explored. Different construction techniques were thus required, and specially developed building machinery and cranes began to change the organisation of the building site;[1] the construction of the Crystal Palace for the Great Exhibition in 1851, for example, required a whole series of different machines, powered by a steam engine.[2]

As a large number of identical or similar parts were essential for making industrial fabrication and serial production economically viable, the number of different building elements was reduced. At the same time their lot size was further increased by using them en masse for the erection of large building volumes. Thus the application of industrially mass-produced building elements first manifested itself in the construction of large-scale steel structures – bridges, train stations, towers, exhibition halls, the glass roofs of galleries and department stores – which were assembled from a manageable amount of columns and beams in available standardised cross-sections. Around the middle of the 19th century, concrete also gained in importance as a result of the industrial production of Portland cement.

By the beginning of the 20th century, the general availability and usability of serial mass-produced building elements in steel and concrete, combined with opportunities to transport them over longer distances, made the use of prefabricated elements much more common even in smaller individual buildings. However, the architectural design of the numerous town houses and office buildings constructed during that time rarely represented the use of industrially produced materials or components, which had somehow simply become state of the art.

Rationalisation and Industrialisation of the Building Process

The building industry started to adopt industrial production methods during the 1920s and 1930s in a push to solve the housing shortage in the growing towns after the First World War. The design of a limited number of identical building elements to construct slightly different housing types aimed to enable serial mass production and time and cost savings akin to those realised in other sectors of industry. At the same time there was an attempt to reduce and simplify the number of stages involved in building on the construction site, to increase the employment of unskilled labour and to shorten the completion time. Walter Gropius's Törten housing estate in Dessau, Germany (1928) is maybe one of the best-known examples,[3] along with the Hausbaumaschine ('House Building Machine') developed and published during the Second World War by Ernst Neufert.[4] This process-oriented initiative differed completely from the approaches of the 19th century in its ambition to change the organisation of the building site instead of just responding to, and borrowing from, the innovations and products developed by other sectors of industry.

The ideas behind the rationalisation of the production of building elements and industrialisation of the building process were first propagated in Europe after the Second Word War, when there was a concerted effort to realise them at a larger scale. For the first time the serial mass-production of elements and use of industrial fabrication methods in the building industry seemed to make sense, because of the tremendous amount of buildings required to meet the urgent task of reconstruction and demand for housing in the postwar period, as well as during the following boom years between 1950 and 1970.

Walter Gropius, Törten housing estate, Dessau, Germany, 1928
The hollow bricks for the wall constructions were serially produced on site and stored until masoned. A crane that moved on rails helped lift heavy building elements.

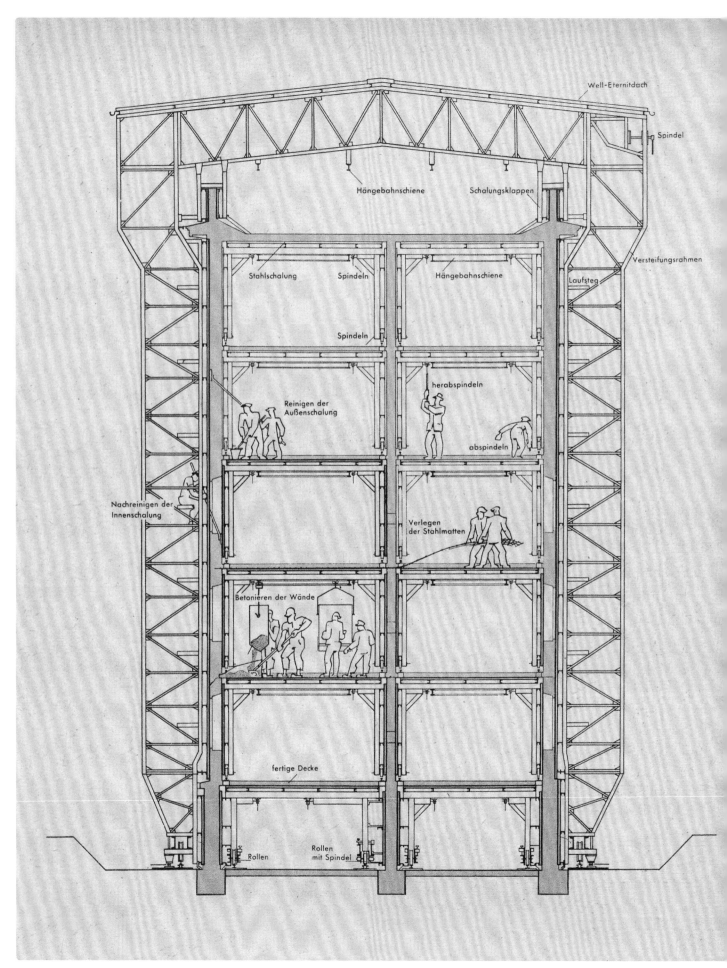

Industrialisation of building: this seems ... to be the keyword, in which direction the building industry has to develop. In our technical age houses should arise like products in a factory ... The engineered work of 'house production' is most distinctively possible in a factory. But also the building site can largely be attuned to it.

— Günther Gottwald, 1951[5]

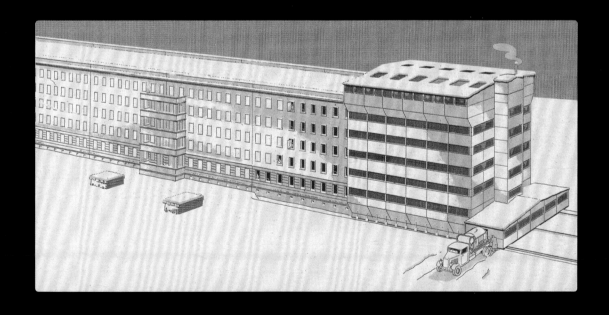

Ernst Neufert, Hausbaumaschine, Berlin, 1943
Section and isometric drawing of Neufert's linear moving 'House Building Machine'. Neufert developed the idea during the Second Word War to solve the housing problem, and published it in his *Bauordnungslehre* of 1942.

While in the 1950s building construction remained quite conventional, with advances largely limited to the use of new materials – plastics, aluminium and composites – as well as larger and stronger machinery, by the 1960s and 1970s a distinct change had come about in the design process. Standardised buildings and building systems were increasingly developed, enabling time and cost savings, as elements were commonly mass produced for the construction of buildings in the housing, education, commercial and industrial sectors. The design of these structures was often subordinate to their production and construction principles, indicating a paradigm shift.

At the beginning of the 1960s, the different elements of large-scale projects often had to be prefabricated in field factories on or near the building site. In the ensuing years, however, an increasing number of independent prefabrication factories began to be built in all industrialised countries. The resulting reduction in the distances travelled to transport materials led to industrially mass-produced building elements being employed at a previously unprecedented scale. The subsequent development and application of different casting techniques, such as slip casting or lift slab constructions, were also significant, as they not only aided the advancement of the industrial prefabrication of elements for later assembly, but also the automation of the building process and the construction site itself.

The 1973 oil crisis, however, proved an impasse for these developments; the Organization of Arab Petroleum Exporting Countries (OAPEC) proclaimed an oil embargo on the US, disrupting the energy supply, triggering recession and unleashing the very real long-term possibility of high oil prices. This created a new awareness of the limits on economic growth, while population rates concurrently stagnated. By the mid-1970s, attempts to fully industrialise the building process declined or were abandoned in both Europe and the US. At the same time, the social problems of towns, which had been planned to be mono-functional, and their mass-housing estates became apparent,[6] and the planning principles of the boom years, based on growth, progress, technology and prosperity, were mostly replaced by ecological strategies and economic considerations. In contrast, Asia did not experience the same dramatic turnaround. Its increasing population and growing cities, resulting from rural depopulation, continued to create demand for the construction of large buildings as well as large building masses. Simultaneously, the lack of skilled labour, especially in Japan, led to the promotion of automation in prefabrication and construction as an alternative to common construction practices.

Staatliches Universitätsneubauamt Marburg, Field factory at the Lahnberge campus, Philipps-Universität Marburg, Germany, 1964
At the beginning of the 1960s, the different elements of large-scale projects often had to be prefabricated in field factories on or near the building site. The building elements for the institutes on the Lahnberge campus of Philipps-Universität Marburg were prefabricated on site and stored until assembly.

Towards Automated Housing Prefabrication

Automation in housing construction started in Japan in the 1960s, with large prefabrication companies such as Sekisui House, Toyota Home and Pana(sonic)Home, which were all descendants of firms that had already successfully employed automation in other sectors. Their manufacturing processes were characterised by a shift away from the construction site to a structured and automated factory-based work environment. In the case of Toyota, for example, 85 per cent of work was pre-executed off site.[7] Nevertheless, the production process in these factories was, for the most part, still conducted by human labour, so it owed more to the organisation of the assembly line than to real automation.

The structured assembly-line work, combined with the advantages of human labour in a factory environment, allowed for the individual adaptation of single parts meeting customer demand without disturbing the production chain.

In contrast to European approaches, where prefabrication was primarily optimised to achieve fast and cheap production of large numbers of identical elements, the major achievement of the Japanese prefabrication industry was their quite early success with customisation and personalisation, as well as their ensuing knowledge of users' demands.[8] The structured assembly-line work, combined with the advantages of human labour in a factory environment, allowed for the individual adaptation of single parts meeting customer demand without disturbing the production chain. They could simply be taken out of the assembly line and replaced manually, to be reworked or finished, before being introduced back into the next stage of the production process, causing minimal disruption to the overall productivity. This approach can be understood as a direct ideational precursor of today's promotion of robotics in architecture – even if it was, of course, far from real automation and levels of productivity a modern industrial robot can achieve.

A common characteristic of the early manufacturing systems of the Asian housing prefabrication industry, which is quite distinct from conventional or traditional product production, was the focus on ongoing development. It was this that optimised them for automated manufacturing. Building systems and manufacturing technologies were mutually adapting to each other.

Misawa Homes, Assembly line quality-control station, Tokyo, 1982
At the last assembly line station, all modules of a prefabricated house pass through a final quality control before they can leave the factory, after which they are packed and prepared for transportation to and assembly on the construction site.

Single-Task Construction Robots

In 1975, after the first experiments in the industrialised prefabrication of 'system houses' (consisting of various structural, exterior or interior wall subsystems etc) were conducted in larger series in Japan, and the first range of products, such as Sekisui M1, the first industrially mass-produced house type, achieved market success, the main building contractor, Shimizu Corporation, set up a research group for construction robots in Tokyo. The intensification of research in this field during the following decade was based on the 1970s 'robot boom' in the general manufacturing industry. The adoption of robots was thus a logical approach for Japanese construction firms.

The single-task construction robots that were subsequently developed were a distinct departure. Rather than merely shifting complexity from the construction site into a structured prefabrication environment, they deployed robotic systems locally on site for demolition, surveying, excavation, paving, tunnelling, concrete transportation and distribution, concrete-slab seeding and finishing, welding and positioning of structural steel members, fire-resistance and paint spraying, inspection and maintenance.[9]

The initial focus was on simple systems that could execute a single, specific construction task in a repetitive manner. Their steering was, in most cases, conducted manually, and was only rarely automated. Since upstream and downstream processes were also not usually integrated in these single-task construction robots, and safety measures were required because of the inferior parallel execution of human work tasks in their operation area, productivity gains were often counterbalanced. The evaluation of the first such robots therefore resulted in the conclusion that an off-site approach would be most suited to the organisation of on-site environments. Sites would be better structured and designed like factories, and the final goal was the implementation of automated manufacturing and construction technologies.[10] Hence research in automated construction was intensified in Japan, leading to the development of so-called integrated automated construction sites.[11]

Thomas Bock, ESPRIT 3 6450 ROCCO (RObotic Computer integrated COnstruction) masonry robot, Karlsruhe University, Germany, 1992–6
opposite: Development of a mobile heavy-duty robot for the construction sector, here applied to robotic on-site assembly.

Taisei Corporation, ROD guiderail for facade painting robot, Shinjuku Center Building, Tokyo, 1988
The guiderail was considered during the design phase of the building in 1976 to enable better robotic maintenance.

Shimizu Corporation, Concrete finishing robot, Kawasaki, Japan, 1987
above top: The robot is deployed on the construction site to finish the concrete floor. Similar to a modern vacuum-cleaning robot, it moves around the floor until it is smooth.

Kajima Corporation, Facade inspection robot, Tokyo, 1988
above: The robot continuously inspects the tiled facade for loose or damaged elements, and is also able to replace them.

Integrated Automated Construction Sites

The first concepts for such structured environments for larger automated construction emerged from 1985 onwards, integrating the earlier single-task construction robots as well as other elementary control and steering technologies as subsystems. These integrated automated construction sites were organised as partly automated, vertically moving on-site factories providing a shelter for on-site assembly, which was controlled, structured and systemised, and unaffected by the weather, as well as for a disassembly process of prefabricated, modular low-, medium- and high-level detailed building components. Robot technology was thus facilitated by the creation of the right conditions to install automated overhead cranes, vision systems and other real-time control equipment.

The conceptual and technological reorientation from single-task construction robots towards integrated automated construction sites was instigated in 1982 by the Waseda Construction Robot Group (WASCOR), which brought together researchers from major Japanese construction and equipment firms in a single initiative. In total, about 30 sites were developed, some as prototypes and others as commercially applied systems. However, their market share and application was limited due to relatively high initial installation costs. These integrated automated sites and their subsystems, such as automated logistics, alignment, welding etc, were thus used mainly when the special conditions (land prices, high labour costs, traffic, noise and waste restrictions) of a project required them.

Since 2008, Japan's major contractors have also developed mechanised and partly automated deconstruction systems, which generally follow the same approach as the automated construction sites, in reverse. The advantage of being able to reduce noise, dust and disturbance of the surrounding environment that these deconstruction systems afford strongly supports contractors' new project acquisition strategies.

Implementation at a Larger Scale?

The historical development of the building industry shows that every innovation in construction technology needs at least one generation to establish itself, no matter how groundbreaking the first experiments or prototypes may have been. While early attempts by the building industry to use industrial materials and production methods were accepted bit by bit (with all their pros and cons) and subsequently changed the organisation of the construction site, the time has perhaps come for automation and robotics to establish themselves in architecture at a larger scale.

Advances in automated construction continue to be developed, especially in Japan and other Asian countries, and are slowly starting to consider the need for customisation of an increasingly individualising society, as well as of the intrinsic conditions of architecture. At the same time, the use of flexible industrial robots in the prefabrication of building elements, as well as in architectural research institutions, is becoming more widespread. However, instead of merely trying to copy and perform long-established construction technologies or prevailing factory automation methods, in order to achieve their inherent performance potential new robotic tools require appropriate conditions, design strategies, kinematics, programming and control. Strong complementarities exist between the actual building – its design, manufacturing and information technology – and its construction and organisation strategy. The next real change will only occur on the construction site once design, management and engineering comply with the robot as a new tool. 1

Obayashi Corporation, Integrated automated construction site concept, Osaka, Japan, 1985
Conceptual idea developed for Obayashi in the mid-1980s. The isometric drawing shows the concept of an upwards-moving factory on top of the building whose elements it is producing and assembling.

Notes
1. Tom F Peters, *Building the Nineteenth Century*, MIT Press (Cambridge, MA), 1996.
2. The Crystal Palace was originally built by Joseph Paxton in Hyde Park, London, for the Great Exhibition of 1851. In 1852 it was deconstructed, and it was rebuilt in Sydenham in 1854. Chup Friemert, *Die Gläserne Arche: Kristallpalast London 1951 und 1954*, Prestel-Verlag (Munich), 1984. For photographic documentation of the later building site at Sydenham see: Philip Henry Delamotte, *Photographic Views of the Progress of the Crystal Palace*, Sydenham (London), 1855.
3. Andreas Schwarting, *Die Siedlung Dessau-Törten. Rationalität als ästhetisches Programm (Logical Reasoning as an Aesthetical Programme)*, Thelem (Dresden), 2010.
4. Ernst Neufert, *Bauordnungslehre*, Verlag Volk und Reich (Berlin), 1943.
5. Originally cited in German in Günther Gottwald, Philipp Stein and Kurt Walz (eds), *Neue Bauweisen: Bildfachbuch Nr 1*, Rödelheim/Frankfurt am Main, 1951, p 3.
6. Interesting in this context is the psychoanalyst Alexander Mitscherlich's pamphlet *The Inhospitality of Our Cities: A Deliberate Provocation*, originally published in German as *Die Unwirtlichkeit unserer Städte: Anstiftung zum Unfrieden*, Suhrkamp (Frankfurt am Main), 1965. Or the discussion about the Pruitt–Igoe urban housing estate in St Louis, Missouri (1954), the demolition of which was referred to as 'the day modern architecture died' in Charles Jencks, *The Language of Postmodern Architecture*, Rizzoli (New York), 1977, p 9.
7. Thomas Linner and Thomas Bock, 'Evolution of Large-Scale Industrialization and Service: Innovation in Japanese Prefabrication Industry', *Journal of Construction Innovation: Information, Process, Management*, 12(2), 2012, pp 156–78.
8. Poorang AE Piroozfar and Frank T Piller (eds), *Mass Customisation and Personalisation in Architecture and Construction*, Routledge (London/New York), 2013.
9. The International Association for Automation and Robotics in Construction, *Robots and Automated Machines in Construction*, Building Research Establishment Ltd (Watford), 1998.
10. Thomas Bock, 'A Study on Robot Oriented Building Construction Systems', dissertation, University of Tokyo, 1989.
11. Thomas Linner, 'Automated and Robotic Construction: Integrated Automated Construction Sites', dissertation, Technical University of Munich, 2013.

Obayashi Corporation, Big Canopy integrated automated construction site, Yachiyo high-rise apartment building, Tokyo, 1997
top left: Construction using Obayashi Corporation's upwards-moving Big Canopy system.

Goyo and Penta Ocean Construction, FACES (Future Automated Construction Efficient System) integrated automated construction site, Tokyo, 1998
top right: The sheltered and structured environment integrates the earlier single-task construction robots as well as other elementary control and steering technologies as subsystems.

Nikken Sekkei, Tokyo Sky Tree, partly automated construction site, Tokyo, 2010
above: The 32-storey, 634-metre (2,080-foot) tall Tokyo Sky Tree radio and television tower in the city centre was built and is operated by Obayashi Corporation.

Text © 2014 John Wiley & Sons Ltd. Images: pp 88-9 © Look and Learn/Peter Jackson Collection/The Bridgeman Art Library; p 90 © Bauhaus-Archiv Berlin and DACS 2013; p 91 © Bauhaus-Archiv Berlin; pp 92-3 © Neufert Foundation, Weimar; p 94 © Staatliches Universitätsneubauamt Marburg, photo courtesy of Helmut Spieker; pp 95-8, 99(tr&l) © Thomas Bock; p 99(b) © Koji Hamada for OBAYASHI Corp., Tokyo, Japan

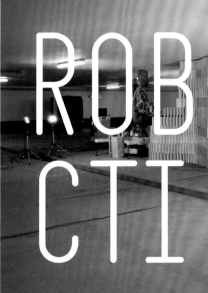

IN-SITU FABRICATION
MOBILE ROBOTIC CONSTRUCTI

Volker Helm

IN SITU FABRICATION

ROBOTIC CONSTRUCTION UNITS ON SITES

Fabio Gramazio and Matthias Kohler, In-Situ Fabrication, Architecture and Digital Fabrication, ETH Zurich, 2012
previous spread: Robotic positioning sequence.

Fabio Gramazio and Matthias Kohler, In-Situ Fabrication, Architecture and Digital Fabrication, ETH Zurich, 2011
below: An 8-metre (26-foot) long modular wall, digitally fabricated in the parking garage of the Department of Architecture at ETH Zurich using an additive assembly method, for which the robot had to reposition itself several times.

Enabling the robotic fabrication of building elements directly on a construction site, in-situ fabrication is a radical and novel approach to the application of digital fabrication in architecture. **Volker Helm** of ETH Zurich describes how an on-site unit, developed with Fabio Gramazio and Matthias Kohler in the Architecture and Digital Fabrication department at the university, offers the first-ever generic robot with applications in a wide range of processes, providing the means of investigating the requirements and capacities involved in non-standardised fabrication.

In-situ fabrication – as opposed to prefabrication on stationary machines – enables the robotic fabrication of building elements directly on a construction site. The application of mobile robotic systems allows for the 'continuous production' and erection of large-format or even complex construction systems over the course of an open-ended, on-site building process. This approach avoids the conventional industrial method of assembling prefabricated elements produced in advance on stationary equipment at off-site or on-site field factories. It also eliminates the need for the often laborious and costly transportation of larger building components to the construction site.

In-situ fabrication requires the employment of an automatic mobile robotic unit that can navigate an unstructured, dynamic construction site, continually detecting and adapting to the surrounding building equipment and components, as well as any material tolerances, imprecision or deviations. A critical necessity is the deployment of a closed-loop system that interconnects the robotic digital fabrication, sensor-driven cognitive skills, and human-machine interaction. This ensures the direct integration of incoming information into the logic and articulation of the building components as well as real-time input that enhances the interlinking of digital data and the concrete material.

In-situ fabrication is a radical and novel approach to the application of digital fabrication in architecture. The robotic unit developed for these purposes with Fabio Gramazio and Matthias Kohler and ETH Zurich, where they share the Chair of Architecture and Digital Fabrication, is the first-ever generic robot with applications in a wide range of processes. Since 2011, it has served as the apparatus by which the requirements and capacities involved in non-standardised fabrication on a construction site are investigated at ETH Zurich.

ON-SITE MACHINES
Since the 1980s, several attempts have been made to employ mobile robotic units on construction sites. Various efforts to develop specialised construction robots were aimed at the semi-, fully automated or remote control of separate construction processes.[1] The most advanced of these were probably the mobile robotic systems ROCCO[2] and BRONCO[3] designed for automated on-site masonry fabrication developed in the 1990s. Architectural considerations were not the prime concern of these early attempts, which were aimed foremost at increasing productivity and economically optimising building processes.[4]

Most of the mobile robotic systems never quite managed the leap from prototype or research to application. This was due in part to the dynamic and unstructured nature of construction sites in comparison to laboratory settings. It can also be attributed to the fact that at that time, technologies were not yet advanced enough to handle the complexity required to develop automated machines capable of operating in unpredictable environments. Consequently, these machines generally ended up being manually operated, serving to facilitate precision in the positioning of heavy building components. It is precisely at this point that the research explorations of in-situ fabrication come into play in the form of a generic robotic unit outfitted with scanning technology that can be employed in complex processes on a construction site.

Unlike mobile construction robots, smaller automated service robots (so-called AGVs – automated guided vehicles) have long ceased to be mere vision in the industry. In logistics they find application in warehouse automation, enabling the efficient transportation and arrangement of shelving systems according to an algorithmic logic.[5] High-end semi- and fully automated units are employed in the military and a current focus in robotic research is aimed at the development of humanoid service robots. In the private sector, inexpensive automatic robotic vacuum cleaners and lawn mowers equipped with basic sensor technology have managed to successfully penetrate the market. The most common application found in the public sector are the inspection and cleaning robots used for clean-up operations following environmental or natural disasters. And robotic couriers are presently being employed in over 200 hospitals worldwide.[6]

Automated robotics research is currently experiencing a quantum leap in development due to rapid advances in hardware technology and increasingly powerful processors. Nevertheless, the realisation of fully automated systems that are capable of handling every conceivable situation is still technically and conceptually extremely challenging and so far only possible to a limited extent. The inclusion of human interaction is therefore recommended for in-situ fabrication when it comes to the simplification of complex processes.

Fabio Gramazio and Matthias Kohler, Stratifications, Architecture and Digital Fabrication, ETH Zurich, 2011
below: The 1:1 Stratifications installation at the FABRICATE 2011 conference in London, performing an experimental demonstration of the handling of building tolerances.

bottom left: The initial configuration of the installation's structure in which differently sized timber blocks are positioned according to a pixel image.

bottom right: The actual configuration after the blocks were robotically assembled, aggregated freely on top of each other.

Fabio Gramazio and Matthias Kohler, In-Situ Fabrication, Architecture and Digital Fabrication, ETH Zurich, 2012
top right: User interaction with the mobile robotic unit whereby the hand gesture is scanned and imported into the CAD software as a line segment.

centre right: The configuration of the wall structure is built accordingly.

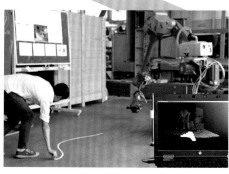

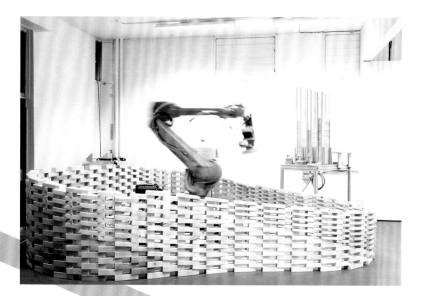

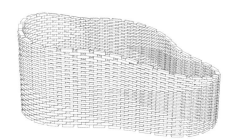

BASIC RESEARCH

Basic research is needed to demonstrate the fundamental advantages of such novel approaches to mobile in-situ fabrication as the continuous digital fabrication of large-format structures. Initial steps have already been taken within the scope of the following research experiments at the ETH Zurich as well as a separate dissertation project.[7] Due to the complex requirements involved in direct fabrication on a construction site, the initial exploration was broken down into several areas of focus: assembling and outfitting a mobile robotic unit; handling building material tolerances; human–machine interaction; and positioning and localisation techniques.

Mobile Robotic Unit

In order to investigate in-situ fabrication, it was first necessary to develop a compact, mobile robotic unit equipped for autonomously handling a variety of construction tasks. Within the scope of a research project funded by the European Union,[8] a mobile system was devised for the first time that could pass through any standard door opening when in a retracted position. Aside from mounting an industrial robot to a track vehicle, additional components such as sensor technology, universal grippers and a vacuum supply were integrated into the robotic system.

Handling Material Tolerances

One of the first robotic in-situ experiments focused on developing and testing a reliable robotic assembly technique for detecting and adapting to material and construction tolerances. Stratifications, a circular test structure, which was first publicly presented at the 2011 FABRICATE conference in London,[9] was assembled from 1,330 discrete wood elements at three varying heights. This experiment employed a robotic unit that could adapt to randomly arising deviations in tolerances. The combination of an integrated sensor (laser rangefinder) and algorithmic operational strategies enabled the robot to self-calibrate in response to any unforeseeable material changes or deviations that arose while stacking the wood elements, enabling it to successfully build a stable structure. In a broader sense, this experiment reveals that the architecture is no longer bound to a purely geometric (CAD) approach.

Human–Machine Interaction

Building on the experiments exploring handling with material tolerances and scanning technologies, a follow-up experiment turned the focus to 3D robotic environment recognition and interaction between the human operator and the robot. Interfaces were designed that would enable a robot to recognise and correctly interpret instructions given by a human operator. The central aim here was to figure out the most basic and intuitive courses of action that would also find application on construction sites. The mobile robotic fabrication unit was outfitted with a 3D scanning device for tracking and processing human hand movements. A 3D camera was then used to record the movement of a hand drawing a freehand line segment on the ground. Once the software had processed the corresponding data input, the robot was able to fabricate rows of masonry bricks along the delineated line. This experiment provided verification that the formation and assembly of building components fabricated by industrial robots can be influenced by human movement.

Fabio Gramazio and Matthias Kohler, In-Situ-Fabrication, Architecture and Digital Fabrication, ETH Zurich, 2009
The experimental setup: an industrial robot, ABB IRB 4600, mounted on a compact mobile track system that is sized to fit through a standard door frame on a construction site.

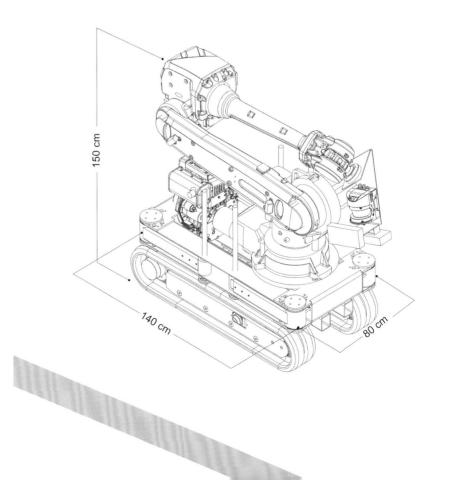

Positioning and Localisation Techniques

The Fragile Structure research experiment served to illustrate the advantage of employing the open-ended in-situ fabrication of a 'continuous building element'.[10] Here, certain operational strategies enabled the automated mobile robotic system to calibrate its own position relative to an existing structure. The experiment was conducted in an underground parking garage where the restricted ceiling height as well as the slanted and uneven ground provided conditions similar to those found on a construction site. The robotic assembly of discrete wooden elements into a complex, non-standardised 8-metre (26-foot) long structure served to further demonstrate the application of in-situ fabrication at an expanded, large-format scale of operation. Following repeated trials using a self-developed local referencing system based on markings and computer vision, it could be verified that after repositioning, the robotic unit is capable of automatically calibrating and building further on existing structures.

FIRST APPLICATION

The current ETH research project, Mesh-Mould,[11] builds on the realisation that although standard industrial robots may be capable of navigating a space with extreme precision, their payload capacity in comparison to objects that are typically conveyed on a construction site is still very limited. Here, the application of in-situ fabrication and its potential for the robotic fabrication of an innovative concrete formwork system is being explored for the first time. In-situ fabrication clearly provides distinct advantages over conventional manufacturing methods for producing complex concrete formwork, since these can only be carried out exclusively with robots. It interlinks the otherwise previously separated fabrication steps – such as the laborious construction of each individual formwork element, the bending and placement of reinforcement bars, and the pouring of concrete – into one continuous digital, spatial production process. This on-site approach to fabrication eliminates the issue of structural weakness in construction joints or size restrictions concerning the transportation of prefabricated components.

FINDINGS AND POSSIBILITIES

Initial key steps have been undertaken to utilise mobile production machines in the production of large-format construction systems in a continuous, open-ended process within unstructured and unpredictable environments. These studies have also provided fresh impetus for conducting further research on this topic in academic and industry settings, as well as finding applications in the construction industry.

Outfitting mobile robotic systems with scanning technology enables an ongoing exchange between the robot and its built environment. It provides robots with the ability to counter-steer and recalibrate in response to material tolerances or construction-site-induced obstacles with the aid of strategic programming. Sensors additionally facilitate interaction between the human and the robot, enabling the enhancement or simplification of otherwise complex operational and software processes.

Future in-situ fabrication requirements are expected to centre primarily on expanding the dynamic behaviour of the mobile robotic unit. Technical and conceptual modifications will enable increased automation and greater sensitivity in handling complex construction processes. The inclusion of multi-robotic cooperation will be crucial to this process, enabling greater dexterity through the application of multiple robotic arms, for example enabling the consolidation of several tasks into one single process. In addition to advancements in hardware and software technology, human–machine interaction and multi-robotic cooperation, future studies will need to focus foremost on scenarios for application in architecture – after all, mobile digital fabrication will not only transform the future building process, but will also have the capacity to significantly change the appearance of our built environment. 𝐷

Sensors additionally facilitate interaction between the human and the robot, enabling the enhancement or simplification of otherwise complex operational and software processes.

Fabio Gramazio and Matthias Kohler, In-Situ Fabrication, Architecture and Digital Fabrication, ETH Zurich, 2012
opposite bottom: The continuous production process of the mobile machinery. A scanning mechanism – developed for finding the centre point of a metal disc and setting it as the origin of the plane – is defined in the CAD model for each new position of the robot.

Fabio Gramazio and Matthias Kohler, Mesh-Mould, Architecture and Digital Fabrication, ETH Zurich, 2013

Large 1.8-metre (5-foot-11-inch) prototype extruded with the mobile fabrication unit. Simulation of a potential application on the building site.

Notes
1. Thomas Bock, 'Construction Robotics', *Autonomous Robots*, 22(3), 2007, pp 201–9.
2. Jürgen Andres, Thomas Bock, Friedrich Gebhart and Werner Steck, 'First Results of the Development of the Masonry Robot System ROCCO: a Fault Tolerant Assembly Tool', *Proceedings of the 11th International Symposium on Automation and Robotics in Construction (ISARC)*, Brighton, 1994, pp 87–93.
3. G Pritschow, J Kurz, Th Fessele and F Scheurer, 'Robotic On-Site Construction Of Masonry', *Proceedings of the 15th International Symposium on Automation and Robotics in Construction (ISARC)*, Munich, 1998, pp 55–64.
4. Tobias Bonwetsch, Fabio Gramazio and Matthias Kohler, 'Digitally Fabricating Non-Standardised Brick Walls', *Manubuild: Proceedings of the 1st International Conference*, Rotterdam, 2007, pp 191–6.
5. Erico Guizzo, 'Three Engineers, Hundreds of Robots, One Warehouse', *Spectrum*, 45(7), 2008, pp 26–34.
6. Richard Bloss, 'Mobile Hospital Robots Cure Numerous Logistic Needs', *Industrial Robot: An International Journal*, 38(6), 2011, pp 567–71.
7. The 'In Situ Robotic Fabrication: Robot-Based Construction Processes on Site' doctoral project is based on a collaboration between the research areas of experimental computer science and architectural digital fabrication. Doctoral candidate: Volker Helm. Supervision: Professor Dr Georg Trogemann, Laboratory for Experimental Computer Science, Academy of Media Arts Cologne, and Professor Matthias Kohler, Architecture and Digital Fabrication, ETH Zurich.
8. The In-Situ Fabrication research work was supported by the EU Project ECHORD. Team: Volker Helm, Dr Ralph Bärtschi, Tobias Bonwetsch, Selen Ercan, Ryan Luke Johns and Dominik Weber. Industrial partner: Bachmann Engineering AG (Zofingen, Switzerland).
9. The Stratifications project was developed in 2011 at the FABRICATE conference in London. Team: Andrea Kondziela, Volker Helm, Ralph Bärtschi and Dominik Weber with Bachmann Engineering AG (Zofingen, Switzerland).
10. The Fragile Structure project was developed in 2012 during an elective course at ETH Zurich. Team: Luka Piskorec, Volker Helm, Selen Ercan and Thomas Cadalbert. Students: Leyla Ilman, David Jenny, Michi Keller and Beat Lüdi. Sponsors: Schilliger Holz AG.
11. See Norman Hack and Willi Viktor Lauer's 'Mesh-Mould' article on pp 44–53 of this issue of △.

Text © 2014 John Wiley & Sons Ltd. Images © Gramazio & Kohler, Architecture and Digital Fabrication, ETH Zurich

Neri Oxman   Jorge Duro-Royo   Steven Keating   Ben Peters   Elizabeth Tsai

TOWARDS ROBOTIC SWARM PRINTING

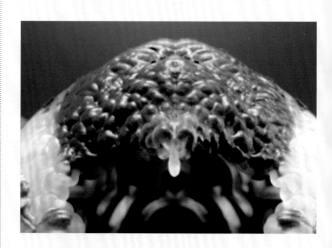

The Massachusetts Institute of Technology (MIT) Mediated Matter Group is honing its research into robotic swarm printing by focusing its efforts on material sophistication, or 'tunability', and communication or coordination between fabrication units. Here, the group's **Neri Oxman, Jorge Duro-Royo, Steven Keating, Ben Peters and Elizabeth Tsai** illustrate this by describing three case studies that investigate robotically controlled additive fabrication at architectural scales.

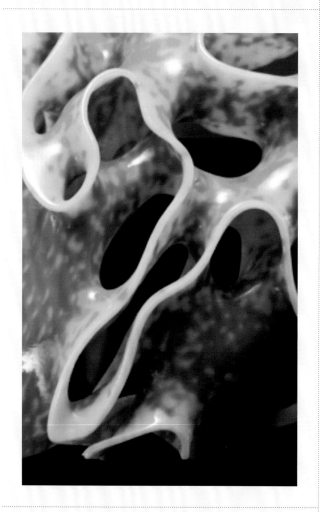

Neri Oxman, Digital Materials, Shock-Absorbing Flexible Helmet and Lung Armour, Centre Pompidou, Paris, France, 2012
above, top and right: 3D-printed structures based on a bitmap printing approach developed in collaboration with The Mathworks, R&D Stratasys Ltd and Professor W Craig Carter of the Department of Materials Science and Engineering at MIT.

Architectural design and fabrication methods have historically evolved in tandem. From adobe brick construction dating before 7500 BC to today's 3D printing technologies, fabrication techniques have developed alongside design strategies and architectural styles. New design expressions have advanced innovation in construction techniques, while new fabrication technologies have inspired designers and architects to further push the envelope of design. This historical perspective allows us to distinguish between fabrication technologies that merely make the construction process more efficient, and others that fundamentally transform our way of thinking about building and buildings.

Today, robotic construction methods have the potential to usher in the next era of architectural design. To realise this potential we must question the basic premises of buildings themselves. Will robotic construction merely emulate manual construction, or become the catalyser for novel building processes? The latter approach invites the designer to consider the ways by which to construct a design process to be as meaningful as the product itself.

Additive manufacturing, or 3D printing, is the process by which to fabricate three-dimensional structures from digital files. Successive layers of material are deposited according to predetermined tool paths until the final form is completed. This fabrication method can be classified with respect to two basic attributes: firstly the degree of material sophistication, also known as material 'tunability'; and secondly the level of communication and coordination between fabrication units. To better understand this distinction, consider two unique structures found in nature: a silkworm cocoon and a termite mound. The silk cocoon represents a highly sophisticated material architecture 'designed' by a single organism (the silkworm) where the mechanical properties of the silk vary significantly from the outer stiffer to the inner softer shell. The termite mound, however, is composed of primitive material with little or no tunability designed by a highly social community of termites. These two attributes – the level of material sophistication achieved through the ability to control physical properties, and the level of communication or coordination between fabrication 'nodes' (termites), can be found and mirrored in the world of digital fabrication.

Manufacturing paradigms to date have been confined to one of these attribute axes, with certain approaches utilising sophisticated tailorable materials, but having limited degrees of freedom and virtually no communication (the silkworm case), and others assembling simple building blocks or prefabricated components in a cooperative fashion with high levels of intercommunication between fabrication nodes (the termite case). The research at the Massachusetts Institute of Technology (MIT) Mediated Matter group aims to combine high levels of communication within and across robotic platforms with high levels of material tunability.

The three case studies here investigate robotically controlled additive fabrication at architectural scales, each representing a unique combination of properties relating to material tunability and communication. First, a robotically layered concrete formwork combining low levels of material tunability with low levels of communication; second, a cable-suspended self-measuring robotic foam-printing platform combining low levels of material tunability with high levels of communication; and third, a templated swarm silk deposition system representing high levels of material tunability with low levels of communication. The article concludes with a brief look at the Mediated Matter group's planned future research into systems combining high levels of communication with high levels of material tunability and the promise of swarm printing as a new direction in architectural-scale additive manufacturing.

Since the mid-1980s, single-node additive rapid fabrication and rapid manufacturing technologies have emerged as promising platforms for building construction automation at the product scale, but with limited applications at the architectural design and building scales. Characteristic of such technologies are the use of mostly non-structural materials with homogeneous properties, the limitation of product size relative to gantry size, and the layer-by-layer nature of the fabrication process. For these reasons, such methods cannot easily be scaled to large architectural systems. However, the ability to additively fabricate at large scales using robotic platforms can help overcome these limitations. This will involve control of material property and variation (material tunability) as well as establishing sufficient communication within and across fabrication nodes (decentralised robotic fabrication) in the robotic construction of large-scale systems.

Material Tunability in Additive Fabrication

To date, additive fabrication systems have acted typically as assemblers of prefabricated components. In general, such low-level subassemblies are structurally componentised and materially homogeneous. The established approach of constructing pre-manufactured building components stands in contrast to the potential of robotic additive systems to deliver highly customised structural and material forms able to potentially adapt and respond to environmental pressures. The Mediated Matter group's research has concentrated on developing variable-property printing platforms delivered through single-node fabrication, focusing on high levels of material tunability.[1] An example of the type of structures that

Neri Oxman and Steven Keating, Functionally Graded Printing, Concrete Structural Experiments, Mediated Matter group, MIT Media Lab, 2010–12
Fabricated linear (left) and radial (right) density gradients in concrete samples allow for material distribution to match stress curves. Samples produced in collaboration with Timothy Cooke and John Fernandez of the Building Technology Program at MIT.

can be achieved here is shown in multi-material 3D prints where the different colours denote functionally graded mechanical properties in 16-micron resolution.

The Mediated Matter group has investigated variable-property printing at multiple scales. Past work developed functionally graded concrete deposition utilising density gradients to reduce mass and improve structural capabilities. Completed work has demonstrated material reductions of between 9 and 13 per cent of the overall mass while maintaining equivalent structural capacity of a fully dense member using radial density gradients in concrete bending samples.[2]

Cross-Platform Communication and Coordination in Additive Fabrication

Progress in swarm construction has typically occurred in the development of sophisticated communication and control protocols to support automated assemblies of basic pre-shaped building components manipulated in

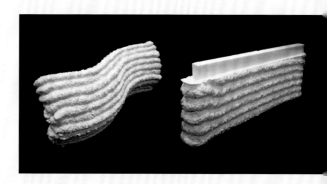

The urethane 3D printing system can print doubly curved structures (left), and the milling effector can be used to subtractively finish the foam according to a digital design (right). The techniques can be robotically combined in a compound fashion to allow for the benefits of both manufacturing approaches without a need for additional fixturing.

predefined paths.[3] Deterministic and stochastic approaches in swarm construction have their merits and limitations: deterministic models offer top-down communication templates enabling robustness and reducing error, while stochastic approaches offer bottom-up intelligence providing responsive and adaptive error control in real time.[4] The goal of the MIT group design research is the integration of these two strategies to achieve top-down control of large structures combined with bottom-up manipulation of localised material features.

Robotic Additive Fabrication Case Study: Print-in-Place

Print-in-Place technology developed by the Mediated Matter group addresses robotically layered concrete formwork combining low levels of material tunability with low levels of communication. It is a construction method utilising fast-curing polyurethane foam as a leave-in-place thermal insulation formwork for castable structural materials. The formwork is designed to be robotically 3D printed on site to allow rapid, custom and efficient large-scale structures. Preliminary research into large-scale 3D printing using this technique has been promising, showing quantifiable benefits in structural strength, construction time, site safety and economic potential.

The technique can be used to additively manufacture polyurethane foam formwork. The resulting leave-in-place formwork is similar to the current construction technology of insulated concrete forms (ICFs), allowing for easier integration into existing site techniques and codes. The spray-foam polyurethane is an ideal material for large-scale 3D printing due to the fast cure time (under 5 seconds), high volumetric expansion rate (up to 40 times), the ability to print double curvature without support material, and the high insulation value of the material.

Robotic Additive Fabrication Case Study: Cable-Suspended 3D Printing

The SpiderBot and CableBot, both developed by the Mediated Matter group, are cable-suspended robotic 3D printing platforms that explore high levels of material tunability using a single-node fabrication system, and high levels of communication using a multi-node fabrication system,

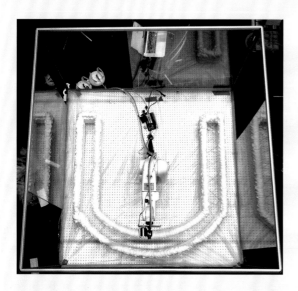

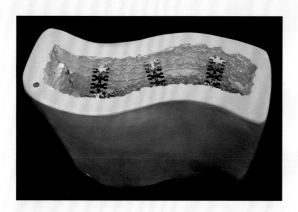

Steven Keating and Neri Oxman, KUKA Digital Fabrication Print-in-Place construction, Mediated Matter group, MIT Media Lab, 2012–13
above top: Additive manufacturing of insulating formwork using a robotic arm and integrated sensor.

above bottom: Layered foam can be printed with double-curvature embedded components (reinforcing bar ties) and finished with robotic processes.

top: A compound fabrication end effector consisting of a 3D printing spray nozzle and a milling effector.

bottom: A milled urethane foam model with the tool paths used to mill the sign.

respectively. The first system explores the technological challenges of material tunability, and the second high levels of communication via an array of robots that share the same environment and deposit a low tunable material.

The SpiderBot is able to build lightweight but large-scale structures, and is capable of rapid deployment in undeveloped terrains. A simple prototype constructed from off-the-shelf winches yields an impressive working volume of nearly 850 cubic metres (30,000 cubic feet). When fitted with canisters of expanding foam, this parallel actuating device is capable of constructing complex building-scale forms via serial layer deposition. The body is composed of a deposition nozzle, a reservoir of material, and parallel winching electric motors.

Print-in-Place technology developed by the Mediated Matter group addresses robotically layered concrete formwork combining low levels of material tunability with low levels of communication.

above: By adding external end effectors to the robotic arm and providing a larger gantry, it is possible to 3D print with acrylonitrile butadiene styrene plastic (upper image) and mill the same piece (lower image) within a single/continuous fabrication process.

Neri Oxman and Ben Peters, SpiderBot cable-suspended 3D printing, Mediated Matter group, MIT Media Lab, 2012

below and bottom: Extrusion head of a SpiderBot prototype – a cable-suspended, large-scale material deposition platform. The diagrams illustrate the comparison between three-axis fabrication motion and a more sophisticated weblike construction motion.

In both the SpiderBot and the CableBot systems, cables from the robot are connected to stable high points, such as large trees or buildings (simulated as a hook system within the installation space in the SpiderBot, and as mounting plates connected to existing structural beams within the installation space in the CableBot). This actuation arrangement allows movement over large distances without the need for more conventional linear guides, much like that of a spider. The systems are easy to set up for mobile projects, and afford sufficient printing resolution and build volume.

Expanding foam or other materials can be deposited by both systems to rapidly create building-scale printed objects. Another material type of interest is the extrusion or spinning of tension elements like rope or cable. With tension elements, unique structures such as bridges or webs can be wrapped, woven or strung around existing environmental or infrastructural features.

The CableBot project uses an array of cable-suspended robots that provides an easily deployable platform from which to print structures that are larger than the spatial envelope of a single robot. The material feed is externalised and reaches the extruder head through hierarchical tubing, in contrast to the SpiderBot where material is carried on the end effector. The motion of every robotic head is set computationally as a rule-based system, exploring the first steps towards the application of swarm intelligence algorithms within the system. The behavioural rules make the robots aware of their envelope dimensions, and of the position of neighbouring robots, so that they can modify their behaviour to avoid collision and hyperextension of the cable system.

The construction strategy explored in the CableBot system is the discrete deposition of soft material drops. This technique enables the emergence of form through robotic node-to-node communication by applying space negotiation rules following each drop deposition. Exploration of this feature is not possible with the continuous layering of material employed in traditional 3D-printing extrusion technologies, such as fused deposition modelling (FDM).

Robotic Additive Fabrication Case Study: Templated Swarm Printing

This project explored a design fabrication approach for robotically controlled additive manufacturing with high levels of material tunability and low levels of communication, inspired by silkworm cocoon construction. It investigates the process of silk deposition by the *Bombyx mori* and

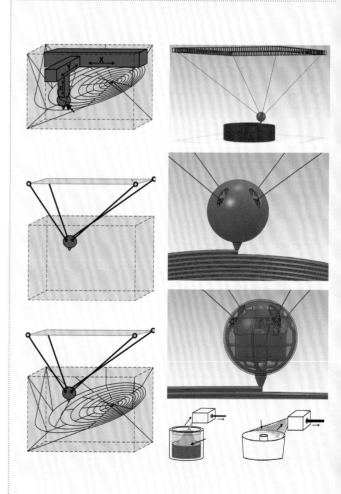

The CableBot project uses an array of cable-suspended robots that provides an easily deployable platform from which to print structures that are larger than the spatial envelope of a single robot.

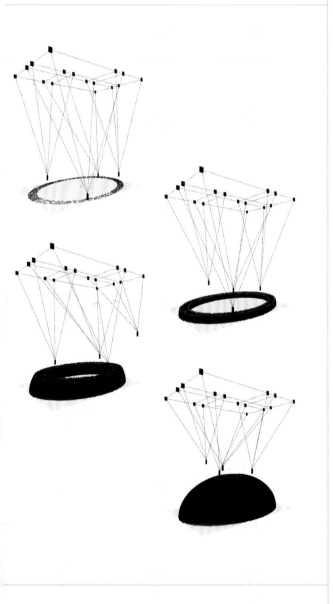

Neri Oxman, Jorge Duro-Royo, Markus Kayser, Jared Laucks and Laia Mogas-Soldevila, CableBot cable-suspended 3D printing, Mediated Matter group, MIT Media Lab, 2013
above and right: A cable-suspended robotic extrusion system for large-scale construction: multiple CableBots simultaneously deposit a dome structure. The mechanical setup employs four timing belts controlled by a system of pulleys that meet at the extrusion head. Designed as compact, lightweight elements, they allow for fast attachment and setup in any environment. The distribution of the robots in space ensures envelope overlap, enabling multiple extruder heads to negotiate material deposition in space and time.

proposes a novel fibre-based templating approach for robotic construction that informs its biological counterpart.

Characterised as 'multi-nodal spinning organisms', with relatively low levels of communication and interaction, silkworms are not bound by social hierarchical structures, but are extremely adaptable to spatial parameters and environmental factors in their immediate surroundings.[5] By studying their spinning behaviour using a magneto-sensor motion-tracking rig, and determining necessary spatial constraints to control it, the system demonstrates that collective multiple silkworm spinning is a viable method for creating a highly tunable fibrous 3D membrane.[6]

In this proof-of-concept experiment, a silkworm 'swarm' was controlled by spatial scaffolding constraints to alter the spinning behaviour of the silkworms from naturally spinning a cocoon to spinning a flat membrane on a template. The superstructure scaffold was digitally constructed using a computational algorithm based on environmental and biological constraints.

Using basic rules such as the silkworm's spinning reach, initial steps towards a digitally controlled system were demonstrated by creating a large-scale (3.6 x 3.6 metre/12 x 12 foot) pavilion deploying a biological swarm. Here, the overall design was controlled and constructed digitally, providing a scaffolding for the organisms to crawl and spin on – 'template printing' – enabling local variations in density and distribution. The superstructure was made of 15,132 metres

Neri Oxman, Markus Kayser, Jared Laucks, Carlos David Gonzalez Uribe and Jorge Duro-Royo, Silk Pavillion templated swarm printing, Mediated Matter group, MIT Media Lab, 2013

top: Motion tracking of the silkworm (*Bombyx mori*) using magnetometer sensors and a 1 x 2 millimetre (0.04 x 0.08 inch) magnet attached to the silkworm. A sample of the captured path data during cocoon spinning is seen on the right.

bottom: Experiments were conducted to determine the relationship between surface area topography and fibre density/organisation produced by the silkworms in different environments.

(49,645 feet) of digitally spun silk thread, while the silkworm swarm deposited approximately 6.5 million metres (21.3 million feet) of silk fibre, creating a highly complex microstructural membrane.

The task of generating a 3D path for digitally fabricating a single non-woven thread, 6.5 million metres in length with highly tunable material properties, is challenging. In the case of the Silk Pavilion, however, this was accomplished by digitally generating only the overall scaffold strategy and leaving the local control and micro-structural fabrication to silkworms controlled through external factors such as changing space configuration, light and temperature. Furthermore, while the geometrical constraints of the space provided the static control factors for the design, environmental conditions such as light and heat provided for dynamic control factors that can enable real-time feedback between existing and desired spinning patterns.

The global design of the pavilion was derived from desired light effects informing variations in material organisation across the surface area of the structure. A season-specific sun-path diagram mapping solar trajectories in space dictated the location, size and density of apertures within the structure in order to lock in rays of natural light entering the pavilion from the south and east elevations, thereby guiding the movement of the silkworms across the structure's surface area.

The Silk Pavilion is an initial proof-of-concept behind the synthesis of digital fabrication and biological swarm construction. The Silk Pavilion experiments are a potential path towards the manipulation of 'biological' builders in achieving further goals in swarm construction and biological fabrication.

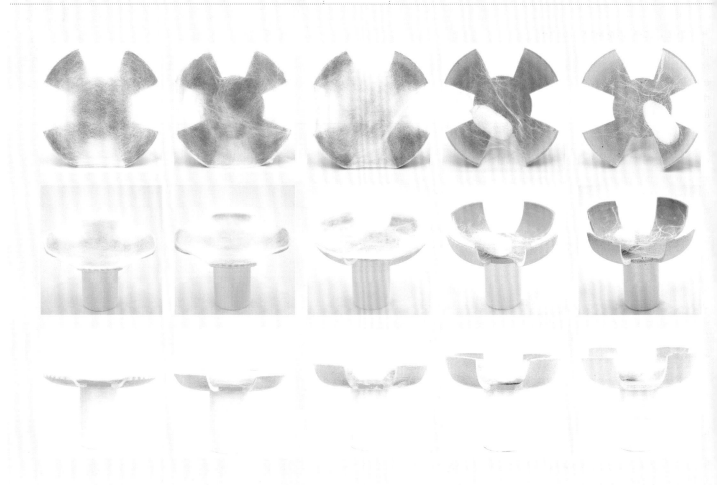

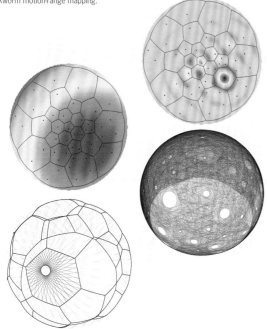

below: A dome structure is transformed into a truncated icosahedron using a customised computation tool that enables 2D fabrication of a 3D structure. Tool paths are generated to enable CNC weaving of the pavilion template. The apertures on the structure are informed by environmental data and silkworm motion-range mapping.

bottom: Silkworms co-spin on a CNC-fabricated silk scaffolding structure to generate the biologically produced pavilion structure.

Towards Robotic Swarm Printing

Robotic (Swarm Printing (RSP) – a robotically controlled multi-nodal additive fabrication platform being explored by the Mediated Matter group – could potentially transform the digital fabrication industry by overcoming the current limitations of additive manufacturing. Material limitations could be overcome by enabling automated construction of structural materials; gantry limitations by enabling the construction of large components made of a collective of cooperative small robots; and, finally, method limitations by enabling digital construction methods that are not limited to layered manufacturing, but also support free-form, woven and aggregate constructions.

RSP combines high levels of material tunability with high levels of communication at construction scales that build on the research discussed earlier in this article. The aim is to create RSP structures that include insulation, structural walls, internal reinforcing bars, tubing and wiring. Using a new approach to large-scale digital construction inspired by the biological world, structures could be designed with integrated and continuous functionalities, and fabricated using a distributed system. At the architectural scale, this would allow for production beyond the limits of traditional construction. Through increased footprint, scalability, robustness, efficiency and material tenability, RSP offers advantages over single-node robotic systems and current robotic building platforms. The MIT Mediated Matter group's future research seeks to unite swarm construction and variable-property additive manufacturing to create complex integrated building systems, inspired by nature, at the micro, product and architectural scales. ⌁

Notes

1. Neri Oxman, 'Variable Property Rapid Prototyping', *Virtual and Physical Prototyping*, 6(1), 2011, pp 3–31; Neri Oxman, 'Structuring Materiality: Design Fabrication of Heterogeneous Materials', *AD The New Structuralism*, July/August (no 4), 2010, pp 78–85; and Neri Oxman, Steven Keating and Elizabeth Tsai, 'Functionally Graded Rapid Prototyping', *Innovative Developments in Virtual and Physical Prototyping: Proceedings of the 5th International Conference on Advanced Research in Virtual and Rapid Prototyping*, Leiria, Portugal, 2011, pp 485–7.
2. Steven Keating and Neri Oxman, 'Compound Fabrication: A Multi-functional Robotic Platform for Digital Design and Fabrication', *Robotics and Computer-Integrated Manufacturing*, 29(6), 2013, pp 439–48.
3. See Chris AC Parker, Hong Zhang and Ronald C Kube, 'Blind Bulldozing: Multiple Robot Nest Construction', *Proceedings of the 2003 IEEE/RSJ International Conference on Intelligent Robots and Systems (IROS)*, Las Vegas, 2003, pp 2010–15, and Justin Werfel, 'Robot Search in 3D Swarm Construction', *First IEEE International Conference on Self-Adaptive and Self-Organizing Systems (SASO)*, Boston, MA, 2007, pp 2013–14.
4. See Ming Lu, Da-peng Wu and Jian-ping Zhang, 'A Particle Swarm Optimization-Based Approach to Tackling Simulation Optimization of Stochastic, Large-Scale and Complex Systems', *Advances in Machine Learning and Cybernetics*, 3930, 2006, pp 528–37, and Christian Jacob and Sebastian von Mammen, 'Swarm Grammars: Growing Dynamic Structures in 3D Agent Spaces', *Digital Creativity*, 18(1), 2007, pp 54–64.
5. Karl von Frisch and Otto von Frisch, *Animal Architecture*, Harcourt Brace Jovanovich (New York), 1974.
6. Neri Oxman, Markus Kayser, Jared Laucks and Michal Firstenberg, 'Robotically Controlled Fiber-Based Manufacturing as Case Study for Biomimetic Digital Fabrication', in Helena Bartolo et al (eds), *Green Design, Materials and Manufacturing Processes*, CRC Press (London), 2013, pp 473–8, and Neri Oxman, Jared Laucks, Markus Kayser, Elizabeth Tsai and Michal Firstenberg, 'Freeform 3D Printing: Towards a Sustainable Approach to Additive Manufacturing', ibid, pp 479–84.

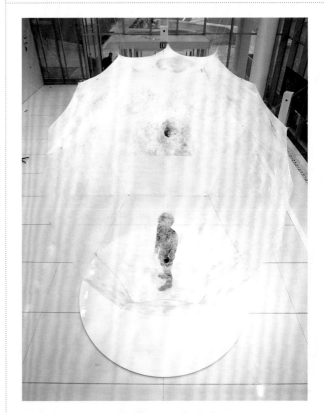

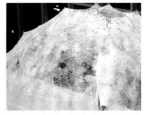

Text © 2014 John Wiley & Sons Ltd. Images: p 108(l) © Neri Oxman, photos by Yoram Reshef for Stratasys; pp 109–15 © Neri Oxman

MACHINES FOR RENT EXPERIMENTS BY NEW-TERRITORIES

Robots have captured the popular imagination like no other technology, fuelling science-fiction narratives in novels, art and film. Here, in a text by **François Roche and Camille Lacadée** of New-Territories, 'Machines for Rent' are explored as a phenomenon that inhabits its own visionary realm, becoming a fertile field for experimentation. These fantastical mechanical creatures are visualised in exquisite drawings by François Roche and Stephan Henrich of New-Territories.

There are many machines, so many desirable machines that in fact pretend to do more than they are doing. In the pursuit of pataphysics – the branch of philosophy that deals with the imaginary realm – they never reveal their deep natures: whether it is their lineage or their illusionary appearance, their genuine qualities or their sham features. Simultaneously speculative, fictional and accurately and efficiently productive, these machines navigate the world of *Yestertomorrowday*, with happiness and innocence, walking briskly through the mountains of rubbish of the 21st century and beyond. These pataphysical machines articulate symmetrically – through weird apparatuses – different arrows of time, different layers of knowledge, but more efficiently they negotiate the endless limits of what we could consider the territory of absurdity, where illogical behaviour is protocolised with an extreme logic of emerging design and geometry, where input and output are described by mathematical rules …

Neither a satire of 'this and other worlds', nor a techno-pessimism or a techno-derision, these machines reside at the very limits of the dystopian or they constitute the limit between the territory of conventions, of certainties and stabilities, where it is comfortable to consider everything legitimated by an order, or an intuition of an order, and by other territories; all the other paranoid, phantasm-like imaginings reported back by travellers.

In a casual and basic sense, machines have always been used to elaborate technicism as the extension of the hand, through its replacement, its improving, its acceleration of the speed and powers of transformation, of production. However, it seems very naive to reduce the machine to this obvious objective dimension, in a purely functional and mechanical approach; limiting it exclusively to a Cartesian notion of productive power, located in the visible spectrum of appearance and fact. In parallel, machines are producing artefacts, assemblages, multiple associations and desires, and are infiltrating the very raison d'être of our own bodies and minds that are codependent on our own biotopes or habitats. Fundamentally, everywhere in nature, at the origin of all exchange processes, in the transaction of any substances, they are the guarantee of its vitalism. Machines' coexistence with nature renders them in effect a paradigm of the body. This is true of all processes, protocols and apparatuses, where transitory and transactional substances constitute and affect simultaneously all the species, where machines' identities and outputs are both object and subject.

In pursuit of this polyphonic approach, we cannot pass over the notion of 'the bachelor machine' as a tentative attempt to integrate mechanical apparatuses in a narrative transaction and transmutation (in the mode of the alchemist). This is the opposite approach to a headlong critique or denouncement of capitalism that highlights the substitution of craftsmen with unskilled workers manning machines (the natural consequence of this now being a mechanical system without workers). Walter Benjamin described this shift as a move from the singularity of production to mass production. This contrasts with the nostalgic romanticism that bachelor machines evoke through our fascination with their sophisticated 'human-made-like' construction: their eroticism or even barbaric eroticism. The 'impulsive urge' and gut-wrenching repulsion they generate means they exist in a permanent state of schizophrenia, vacillating between the simultaneous potential of production and destruction. Both positive and negative processes are the product of the same industrial system; their genesis is consubstantial, and their collateral effect diametrically in opposition. They are both dependent on this schizoid potential.

Following are a few examples of those pathological strategies of narration-production.

DARWINIAN STAR-GATE

Instructions

Stand up and face the ghosts in the depths of your private garden!

Rent this vehicle to transport yourself from a seated, peaceful, sleepy, archaic body posture to a standing, lucid, awakened position that induces bravery in those faced with the present.

Powered by photovoltaic cells, the Darwinian Star-gate's arms unfold on their way from a panoptical to a worrisome heterotopic space you would normally refuse to see.

The star-gate machine introduces the passage of time between two constructions of different origins and periods. As a strategy for questioning the orientation of the arrow of time, it is able to quieten the anxiety of misunderstanding provoked by the shifts between the Modern to Postmodern, Postmodern to Digital, and Digital to Robotic Computational extension, in a 'beam me up, Scotty' rhyzomatic shortcut. The travel could take an evolutionary and/or regressive trajectory. But mainly this machine is most efficiently used as a vector of discovery that reaches a point of uncertainties, of un-determinism, to escape from a zone where everything has already been flattened, classified and validated.

Its first use and development was for the 'broomwitch' experiment.

Precautions for use

Using the vehicle too often might cause a sensation of time deprivation and sometimes immortality, but also ultimately provide a good excuse for denial of your duties in a given time. You might also lose the sense of time passing, which can significantly impair synchronising motor actions. Abuse of the device can be extremely dangerous for mental health and seriously affect the user's temporal perception, especially in regards to the notion of a specious present. Ultimately, it can cause memory loss.

On the contrary, overexposure to the present time (staying in one time or another) might cause the user depression, cynical behaviour or other pathological distress, which the vehicle shall not be held responsible for …

The device does not work for French architecture, which already confuses its origins.

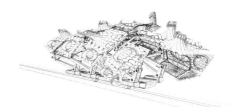

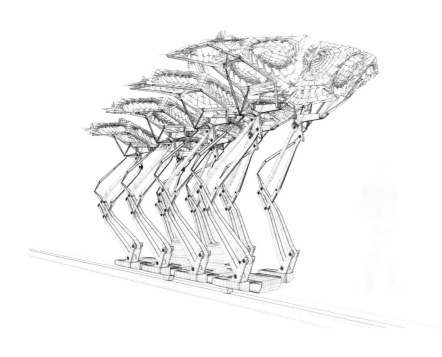

New-Territories/François Roche-Stephan Henrich),
Darwinian Star-gate, 'broomwitch', 2009

ANTIPERSONNEL NYMPHOMANIAC WANDERER

Instructions

Rent this machine to brave the danger and bring back to you on her back rotten species, decomposed biomass, from any 'no-man's land'.

The Wanderer can be transformed for collecting other material. All robot 'tuning' of terminations, articulated arms, legs and tips is authorised, on the condition that you return the machine in its initial state.

The machine collects any ingredients to be recycled in a new productive use. This grants a second life to the waste, and the trash in polluted areas such as post-military zones with unreachable infrastructure interstices.

Legends and fairytales are simultaneously transported out of the deepness of those abandoned situations, as in a 'Stalker' experiment to touch the unknown. Please beware of the backlash of those creatures.

Its first use and development was for the 'he shot me down' experiment.

Precautions for use

The machine is originally built with a very high self-estimation sensorial device, as well as a danger-blinding component, both necessary for its brave actions and responsibilities. However, depending on the environment it is exposed to, the machine could be subject to sudden and violent changes in self-esteem.

In case of failure or minor breakdown (if the danger-blind component gets hurt), the machine will exhaust itself until suicide. If you notice that the machine repeats a very high exposure to dangerous situations, switch it off to avoid risks of suicidal tendencies disguised as bravery.

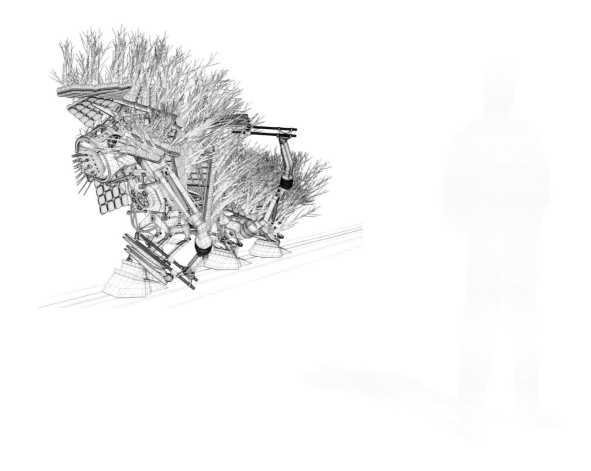

New-Territories/François Roche-Stephan Henrich, Antipersonnel Nymphomaniac Wanderer, 'he shot me down', 2009

INTROVERTED ECZEMETAL RECYCLER

Instructions

Transforming informal heaps into deformed ones, this machine recycles waste from metallurgic and construction sites into potential troglodyte morphologies.

This machine is still in development. It is thus available for rent under a special discount as a beta test. The provision of sufficient energy levels for the effective gathering of steel has not yet been accurately gauged, and dysfunctions may easily occur when the Recycler is in operation. We recommend for this machine only to be rented in parallel with the crane that is able to stabilise its agenda and positioning. We require feedback from customers to improve the reasons behind its design, which appear for now weak. This machine will be removed from the catalogue if there are no further reasons for it being in existence.

Precautions for use

Due to its lack of resolution, this machine is especially vulnerable.

Protect it from the feeling of identity loss by engaging with it on a private level – otherwise it might show a tendency to confuse its own being with the built environment, and develop skin camouflage diseases in order to disappear inside its own construction. An early stage of depersonalisation can be spotted by its tendency towards metallic somatisation.

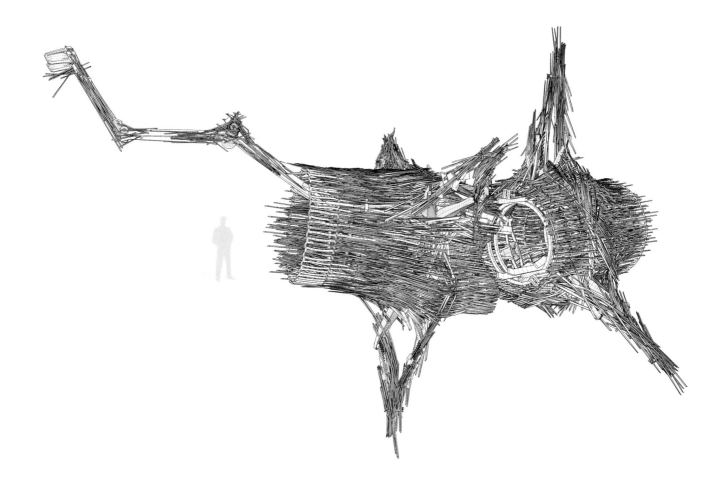

New-Territories/François Roche-Stephan Henrich, Adam Orlinski and LAB Angewandte, Introverted Eczemetal Recycler, 2009

DIFFERENCE AND REPETITION/INTRICATE RANDOMISER

Instructions

Rent this device to populate a surface that will be revealed by the trajectory you convey by impulse to the machine. Its multiple arms will follow a dance of intricacy in compulsive articulated movements, giving ideal programmable empirical shape and outcomes.

This machine has to be rented with a specific number of components (only available in packages of 500 units) to be populated in any condition, any situation. The individual component is developed as a Velcro termination, self-attached by a comb-feather design, with variable positions able to assume, at your convenience, polyphonic structures; be they massive, fluid, opaque or transparent.

The machine is able to be packed in a pick-up of 3 x 2 x 1 metres (10 x 7 x 3 feet) including the tracks that are 10 metres (33 feet) long. Please refer to the installation instructions for ascertaining the dual positions of the machine/component on the ground. You will be trained in the inverse cinematic process that will enable you first to draw the structure manually, as curves in space by manipulating the machine tips, and secondly to discover how the footprint of your handy movement is becoming the trajectory of the components stacking, automatically repeated and assembled by the machine (4 metres/13 feet high maximum). The intricate packing fabrication will follow the isocurves you defined in the space in a repetitive adaptation.

Precautions for use

Due to the requirement of unpredictability of its work, the machine is subject to bipolar disorder, alternating manic, hypomanic and depressive episodes of varying lengths. Although these episodes are necessary to the nature of the random intricacy process, they might in the long run cause side effects such as racing thoughts and rest (OFF mode) deprivation.

Beware of possible exhaustion of the machine, as well as of the feeling of impuissance before its never-ending chore.

On the contrary, if the machine shows repetitive, ordered or systematic combination processes, bring it back to the shop immediately for emergency reprogramming.

New-Territories Academia/François Roche-Stephan Henrich, Guo & Wang and LAB USC, Difference and Repetition/Intricate Randomiser, 2010

BODY-BUILDER-SHITTER

Instructions

Rent an agile hyper-proteined device, shitting liquid concrete in a vertical phallic extrusion, which is turned into coagulations that it stands on to continue the construction process in defiance of gravity.

The Shitter is only made available to rent to a minimum of 30 families, dedicated and driven by a bio-politic decision.

The device is a usable, operative machine for a self-organised micro-urbanism conditioned by a bottom-up system. The 30-plus families, called 'the multitude', are able to drive the entropy of their own system of construction, their own system of '*vivre ensemble*'. It is based on the potential offered by contemporary bioscience, the rereading of human corporality in terms of physiology and chemical balance to make palpable and perceptible the emotional transactions of the 'animal body', the headless body, the body's chemistry, and information about individuals' adaptation, sympathy, empathy and conflict when confronted with a particular situation and environment. The construction process developed through 'machinism' – indeterminate and unpredictable behaviour – with the creation of a secretive and weaving machine that can generate a vertical structure by means of extrusion and sintering (full-size 3D printing) using a hybrid raw material (a bio-plastic cement) that chemically agglomerates to physically constitute the computational trajectories. This structural calligraphy works like a machinist stereotomy composed of successive geometrics according to a strategy of permanent production of anomalies: with no standardisation, no repetition, except for the procedures and protocols at the base of this technoid slum's emergence.

Its first use and development was for the 'an architecture des humeurs' experiment.

Precautions for use

The machine is set in between anal and foecal stages, leaving both unresolved in order to achieve full development of its construction capacities. Anal expulsive behaviours, as well as exhibitionism, are frequent phenomena of the machine and are to be considered as signs of good health.

Placed in an extremely social zone, these behaviours could later develop into paraphilia: manifesting in hyperbolic intensifications, distortions, monstrous fruits of erotic expression outside of normal eroticism. It is strongly recommended, therefore, not to place it in public zones (ie outside of your own multitude).

The device is also slightly narcissistic, which could provoke strong reactions in similar devices of different multitudes.

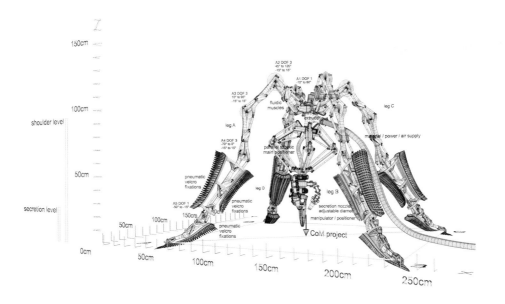

New-Territories/François Roche-Stephan Henrich, Body-Builder-Shitter, 'An architecture des humeurs', 2011

OCD PACKER

Instructions

Rent this extremely efficient packing, ordering, classifying, numerating and xyz-positioning machine, for an endlessnessless stacking and staggering. The Packer is only available for long-term rent.

The machine works to extend existing construction, by testing the possibility of wrapping, smearing and invading a previous situation to develop a surrounding maze with multiple uncertain trajectories and '*parcours*'. The morphological trap it creates is both a jail and a protection apparatus. This dual strategy avoids the occupant perceiving their own madness and protects others from their own pathologies.

Participants require a personal agreement and discharge to play this game as a 'voluntary prisoner', lost in the permanent entropy of packing. In any case you could use, if necessary, RFIDs on PDAs to rediscover positioning – but at your own risk.

Its first use and development was for the 'Olzweg' experiment.

Precautions for use

In order to achieve high efficiencies in ordering, numbering, arranging, checking, cleaning, etc, the machine was implanted with intrusive thoughts that can produce uneasiness, apprehension, fear and worry.

The repetitive behaviours aimed at reducing these anxieties can also manifest in an aversion to particular numbers or in the absurd repetition of nervous rituals.

If you notice such signs of obsessive compulsive disorder, please bring the machine back to the shop immediately for a diminution of input anxieties.

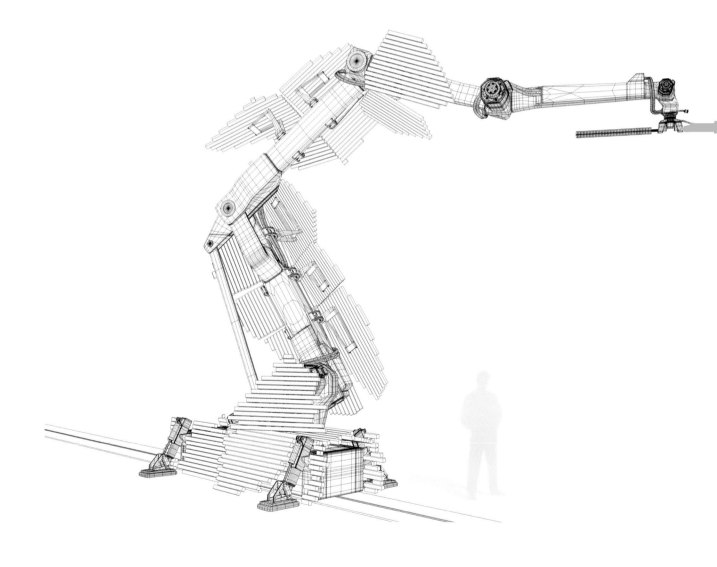

New-Territories/François Roche-Stephan Henrich,
OCD Packer, 'Olzweg', 2006

ALGAE-SACHER-CYCLOTHYMIA

Instructions

Rent this under-seawater device that behaves as an extractor removing algae and extracting chemicals (calcite) and particles from the water in order to agglomerate a masochism structure. The progressive accumulation is condemned to be pulled and pushed by the current and tide, which drives the orientation and the progression of the crystallisation without a forecasted positioning agenda.

The machine is usable only in seawater, which contains approximately 400 milligrams per litre of calcium and represents 1.6 tons per cubic kilometre. The calcium is obtained from dissolving rocks such as limestone, marble, calcite, dolomite, gypsum, fluorite and apatite. Before renting you need to request a survey to confirm the quantity of calcium in your location. We can provide this expertise.

In order to function, the device requires a water depth of between 6 metres and 20 metres (20 and 65 feet).

The extraction, transformation processes are patented. The chemistry filtering and reaction cannot be divulgated in these instructions for use. Please do not open the sealed core of the machine; it is toxic.

Precautions for use

The machine is built with a total submission to external factors such as currents, tides and lunar eccentricities. The more it is ill-treated by the water, the better it will work. The machine is also cyclothymic, subject to mood swings, and is voluble in its responses to the water humiliations.

Due to the mixture of these characteristics, the device is susceptible to construct totally useless structures, and cannot be held responsible for the unusable nature of the structures built. You rent it at your own risk.

In extreme cases of maltreatment, where the machine is overexposed to water or other environmental factors, it could become self-defeating or suicidal. Ultimately, it could completely stop functioning.

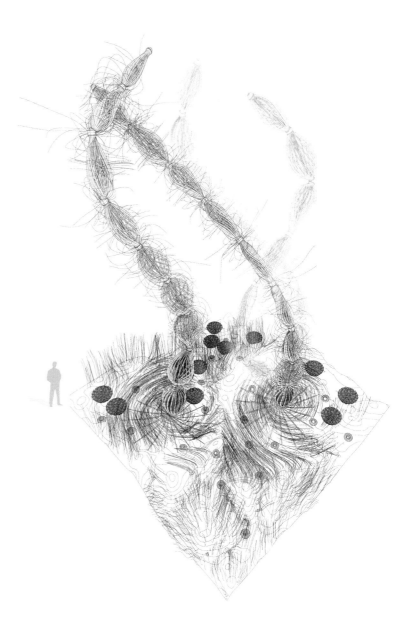

New-Territories Academia/François Roche-Stephan Henrich, LAB Angewandte, Mirko Daneluzzo and Martina Johan, Algae-Sacher-Cyclothymia, 2009

BULIMIC ENCLOSURE-WEAVER

Instructions

Rent this silk cocoon-weaving device – preciously precise and accurate – to create temporary buildings, camp sites, outdoor workshops or garden parties.

Do not complain that this machine is both the producer and the structure of the production, trapped in its own net. It is its own process of know-how.

The silk membrane could be waterproof or not. Please refer to the density of knitting in the machine's instructions. The wire is the product of bio-production, starch and flax. Its lifespan is around 10 days before it degrades and loses its structural resistance. This melting condition is 100 per cent natural, and the process of necrosis will provide nitrogen and nutritional elements to the ground. Do not be afraid of the ostensible pollution it seems to generate.

Different time spans for synthetic silks are available on application.

The synthetic silk wire is provided by a bobbin of 10 kilometres (6 miles).

Precautions for use

The machine is conditioned to have a lack of bodily feeling in its surfaces in order to keep it endlessly weaving surfaces. However, it can unexpectedly reject the surface and return to its body, inducing the formation of a protective cocoon around itself.

This bulimic tendency to re-create a virtual dimension of potential traits, connections, affects, and movements around its own body are symptoms of the future loss of the machine inside its own production.

At this stage nothing should be attempted to stop the claustrophobic process.

New-Territories/François Roche-Stephan Henrich, Qi Su and Shenyuan Guo and LAB USC, Bulimic Enclosure-Weaver, 2009

THE ASTROLABE STUTTERER

Instructions

Rent this device – the Astrolabe Stutterer – to ascertain the level of threat posed by two discrete planets, the sun and the moon, and the human pathologies they produce. This machine detects any potential harm that these planets threaten, and secures your negotiation with the celestial vault: protecting you from magnetic storms and radiation from the sun, and the psycho-lupus affliction of the moon.

The Astrolabe Stutterer cannot simultaneously maintain an equal position between the sun and the moon, except during eclipses, every 6 585.32 days, exactly 18 years, 10 or 11 days and 8 hours, depending on the occurrence of leap years.

The part of the machine dedicated to the sun indicates the planet's celestial cycle. It particularly highlights any gap in the sun's position and the degree of protection afforded by the interaction of solar rays with the ozone layer, which has had its impact depleted by UV emissions. The device can be used in correspondence with uranium powder, which has a natural afterglow that indicates the intensity of UV emissions. The uranium powder is provided with special conditions, because of the emission of alpha rays (below the administrative threshold), which have been agreed by legal settlements. This machine has the potential for a double paranoia: one harmful substance acts as sensor to another, providing a chain of past–present–future industrial collateral effects.

The moon part of the device points to the symptoms of the moon: the forces of attraction, and fear of transformation (real or illusionary). It works as a vector of 'science of the imaginary', through a pataphysical approach. Nothing seems real, but everything in fact affects your metabolism.

Its first use and development was for 'TheBuildingWhichNeverDies' (TBWND) experiment.

Precautions for use

Due to the dual nature of the object to be read, schizophrenic episodes of low intensity are normal and even necessary to the effective functioning of the device.

If placed under ambiguous coordinates, the clock is subject to delusive and paranoiac interpretations of the astral movements and this will induce disorganised reports and drawings, impossible to read or understand under normal circumstances.

Hallucinatory episodes may include the creation of a third aster or the predictions of absurd mortal events such as overexposure to the moon's dangerous UV light or a daytime invasion of werewolves.

If pushed to its extreme, the device will run from purposeless agitation and motions to complete catatonia in which case it is recommended to unplug. ⌂

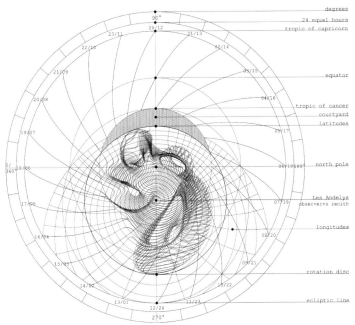

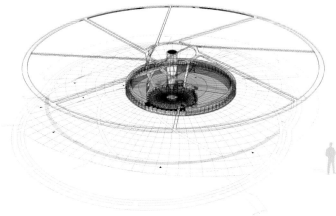

New-Territories/François Roche-Stephan Henrich, The Astrolabe Stutterer, 'TheBuildingWhichNeverDies' (TBWND), 2010–12

Text © 2014 John Wiley & Sons Ltd. Images © New-Territories/François Roche/Stephan Henrich

COUNTERPOINT

Tom Verebes

CRISIS! WHAT CRISIS? RETOOLING FOR MASS MARKETS IN THE 21ST CENTURY

From his base in Southeast Asia, **Tom Verebes**, Associate Professor in the Department of Architecture at the University of Hong Kong (HKU), asks what the hardcore realities are for robotics in architecture in this new 'Asian Century'. At a time when China strives to accommodate 10 million new urban citizens each year by rolling out high-rise residential blocks with standardised serial production techniques, what hope is there that the homogeneous Chinese city will apply robotic tools of mass customisation for the construction of distinctive, rather than generic structures?

The cultural perception of robotics today has its origins in the popular imaginings of science fiction in film and literature. In the 1960s television cartoon *The Jetsons*, robots entered every vestige of society – its manifold domestic gadgets, its infrastructure, its architecture and, most pervasively, its cities. In the future, as it was sketched in the mid-20th century, technology is equated with culture, and in *The Jetsons* culture has conquered all of nature, including gravity. In the science fiction of the latter Modernist era, technology was, as often still is, heralded as a labour-saving set of systems and mechanisms augmenting efficiency, not only for domesticity, but also for all of urbanity. The robot, in 1960s cartoon techno-porn, was humankind's slave without any ethical or moral dilemmas. *The Jetsons* encapsulated how the future was envisaged during the height of the 'American Century'. In the hardcore realities of this new 'Asian Century', if science fiction is still a relevant genre to speculate on the future, how does today's sci-fi compare to America's giddiness half a century ago?

The unprecedented scale, rate and extent of urbanisation in China are of particular concern given China's reliance on standardised production methods. A more poignant issue to be addressed is the homogeneous quality of contemporary Chinese cities.[1] China is currently

The Jetsons (Jane Jetson and Rosie the Robot Maid), 1962–87
Hanna-Barbera's *The Jetsons* television cartoon show depicted a techno-utopia in which robotic technologies permeated all aspects of domestic architecture within an urbanism facilitated by its smart mobilising infrastructure.

in the midst of the largest city-building era ever. As in other periods of industrialisation or rebuilding after wars or natural disasters, China's tremendous project of urbanisation is driven by the dictate of building housing for over 10 million new urban citizens each year, for the foreseeable future. Urbanisation in China, accelerated by its centralised model of planning and investment, and facilitated by basic 20th-century techniques of material realisation in steelwork and concrete, seems unrelenting in its drive to house these vast numbers of new urbanites. Currently more than 75 per cent of the world's urban population lives in Chinese cities, 170 of which have populations of over a million; statistics that are only set to rise in future. The problem is, how will robotics escape the regime of standardisation, and how will investors, rather than construction managers, get on board with the revolutionary changes occurring in post-Fordist custom architecture to proliferate serially at the scale of cities?

For these new powerful tools to diverge from the goals of the 20th century – which fused an unholy bond between Fordism and capitalism – the pursuit of differentiated spaces and systems is paramount. Prototyping methodologies, which include robotics, are increasingly being adopted in the making of non-standard architecture at the scale of discrete iconic buildings, facades and other building systems, interiors and furniture. My concern, however, focuses on application of robotic tools of mass customisation for the construction of distinctive, rather than generic, cities. This

Qilai Shen housing estate, Shanghai, China, 2011
Aerial photo of a housing estate in Shanghai, indicating the extent of standardisation deployed in the making of repetitive housing. The legacy of standardised modes of production is perpetuated in the vast project of urbanisation to house hundreds of millions in Chinese cities.

issue of ⌂ speculates upon the ramifications of robotic production technologies whose applications are accelerating at the innovative fringes of architectural design, practice, teaching and research, yet the potential of robotics remains largely untested at the vast scale of the contemporary city.

Robots literally find their way to architecture via advancements made in parallel design industries, such as the automotive industry. Fast-forwarding to the future, Amazon's and Google's recent research into robots for a delivery service is looking like the imaginings of *The Jetsons*. Architecture has a promiscuous relationship with new technology, and yet cities, while mechanised, are still far from absorbing the impact of robotics. Architecture and cities were understood within Modernism as naturally evolving 'towards the establishment of standards (types, norms), of a homogeneous style'.[2] As a result of the mass production of architecture, cities in the 20th century began to take on repetitious qualities and increasingly share characteristics. This perpetuated a generic, Modernist approach to city formation and expansion that, in terms of their spatial and material organisation, led to the creation and development of monotonous and overly standardised cities.[3] In China's current housing boom, the legacy of such standardisation in mass-production methods is manifested in low-end on-site manual labour, which may prove to be the entry level for robotic construction.

This issue of ⌂ forms a growing body of evidence of a paradigm shift from mass production to mass customisation. At

the core of the ambition of this grouping of designers and entrepreneurs lies a critique of ubiquity, universality and monotony of the Modernist industrial paradigm. Cities are inextricably tied to a society's model of production, and the prevalence of generic urbanism, which can be found anywhere and everywhere, comes out of the legacy of globalised Fordist mass production. In contrast, mass customisation, as a model of contemporary production, aims for non-standard effects from simple parts, leading to more complex and compelling architectural entities, and potentially to more unique and distinctive urban environments. However, robotics may not entirely supersede old technologies – as suggested by Kittler's observation on new media: 'New media do not make old media obsolete; they assign them other places in the system'[4] – if its current role remains unclear in terms of scalability, and the design, production, manufacturing, assembly, management and maintenance of our cities. The impact of robotic urbanism may be more than achieving large-scale implications of these new media, and more importantly to facilitate more distinctive and unique cities.

Laboratory Experiments and Entrepreneurial Initiatives

Once classified as 'blue sky' research, robotics targets the frontier of manufacturing, fabrication, construction, assembly and management. A new technological culture is taking shape through the mischievousness of small groups of well-funded researchers in a few niche architecture and engineering schools. The repercussions of small-scale experiments in robotics in this current era of 'Personal Fabrication' may indeed surpass the limitations of the prototyping expertise of a boutique or specialised set of methodological interests of a small demimonde design community. The high-precision engineering methods currently found proliferating small-scale experiments are driven by 'file to factory' flows, or 'bytes to bots'. This research arena is quickly being adopted and applied by savvy, opportunistic entrepreneurs who set up outsourcing robotic fabrication labs, and may indeed be the catalyst for robotics to migrate from the elite to everyday manufacturing. These startups are speculators in an emerging service industry; small, diversified companies 'taking a punt' on a number of small projects in the hope that the technology proliferates; at stake is more than economic gain. Most importantly, they demonstrate how technological innovation has become a form of cultural practice through hotwiring and customising machinery. Ideologically, is it this gang of experimental designers who will bring robotics to the mainstream, with their custom products, small pavilions or, at best, tall facades? Or will entrepreneurial initiatives, such as those by Odico Formwork Robotics, be the catalysers, as the firm states in its mission, 'to transgress architectural robotics from pilot research production to industry mass-adoption' (see pp 66–7)? If the provisional results documented in this issue indeed prove to be scalable in the coming years, will robotics exceed the repeatable production mode of Fordism, namely standardised mass production? Once again, what role will robotics play in shaping the cities of the 21st century?

Mass Markets

No doubt the construction industry will see benefits from automation; not only economic ones, but also in substantiating the potential for greater customisation. At stake are not only the prerogatives of faster and cheaper, but also better and more distinct – very different, rather than all the same. In which ways can automation be used to generate high orders of differentiation rather than perpetuate the legacy of repetitive production of the 20th century? More insidious is the legacy of Fordist industrial modernisation in the homogenised and uniform cities now being produced. Mechanisation and standardisation grew out of the Industrial Revolution, which shifted the work of society from bespoke one-off products towards production for the masses. The notion of a complex 'piece' of architecture being beyond the scope of manual material manufacturing is now widely accepted, given the near-countless examples of large-scale architectural projects that demonstrate the potential of computational design software perhaps more than that of computer-aided manufacturing. Jumping scale seems a precondition for robotics to migrate from boutique production to production for mass markets.

Fabio Gramazio's and Matthias Kohler's work at ETH Zurich, with their expert team of researchers and students at the Future Cities Laboratory (FCL), Singapore-ETH Centre for Global Environmental Sustainability (SEC), looks at how, for example in their *Structural Oscillations* installation at the 2008 Venice Architecture Biennale (see pp 6, 14, 15 and 59), small-scale prototype walls can be scaled up to address issues associated with the tectonics of high-rise tower structures, spaces and facades. Their design research confronts the scales of production inherent to mass production: 'It is no longer the delayed modernisation of an industry we are witnessing, but rather a historical departure: the modern division between intellectual work and manual production, between design and realisation, is being rendered obsolete' (see their 'Introduction' to this issue on pp 14–21). However, despite the innovative provisional achievements of their research, robotic construction has not as yet had a substantial and pervasive impact on the conventional construction processes that create the architecture of today's cities.

Despite the increasing normalisation of human and machine interaction, in-situ, bringing greater precision to the inherently inaccurate building site, robotics are but one set of technologies able to address the legacy of repetitive construction. The promoters of such technologies may also have more robust arguments than the economic benefits of reducing time and cost. But aside from these efficiencies, how will on-site robotics propel current building culture to define itself differently from the past? How will robots create possibilities for non-standard cities, rather than efficient building programmes, and convince project managers and client's agents of their viability to be integrated as new conventions. The 'game changer' for construction will be a departure from the controlled laboratory environment of interior spaces in which the prefabrication of architectural elements takes place, to the exterior construction site, or 'mobile in-situ fabrication', where a plethora of parallel processes occur in the field. Volker Helm's sci-fi forecast (see pp 100–107) is for mobile robots to roam cybernetic construction sites, on earth, as it were, where their behaviour is controlled through intelligence in order that they can interact, as happy machinic co-workers, with their human inter-actor construction workers.

When Henry Ford invented and deployed the production techniques of the assembly line, was he accused of promoting an isolated and niche mode of production for an elite market of products? Despite the large unit numbers in which the Model T car was produced, and the invention of a new mode of production for the masses, lest we be reminded how, in

The paradigm of mass customisation offers clues as to how to target non-standard design while responding to the contingencies of rapid and large-scale urbanisation. But this view is again biased towards regions such as East Asia that are experiencing the most widespread extent and scale of urbanisation.

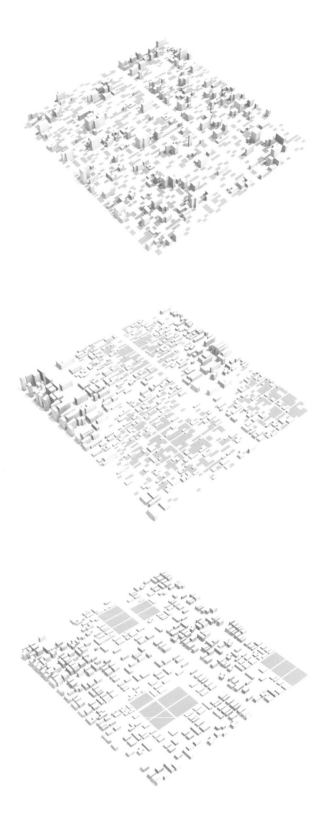

dotA & OCEAN CN, Yan Jiao Hua Run 4D City, Hebei Province, China, 2011
In this masterplanning project, several schemes were presented as low-, medium- and high-density urban models, demonstrating the potential for customisation of urban spaces.

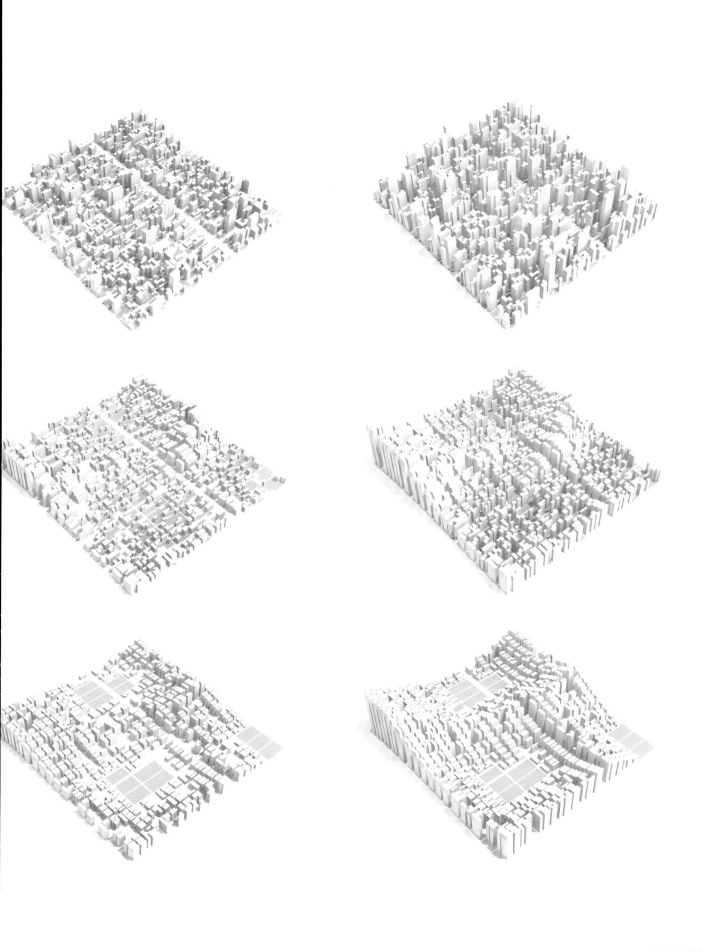

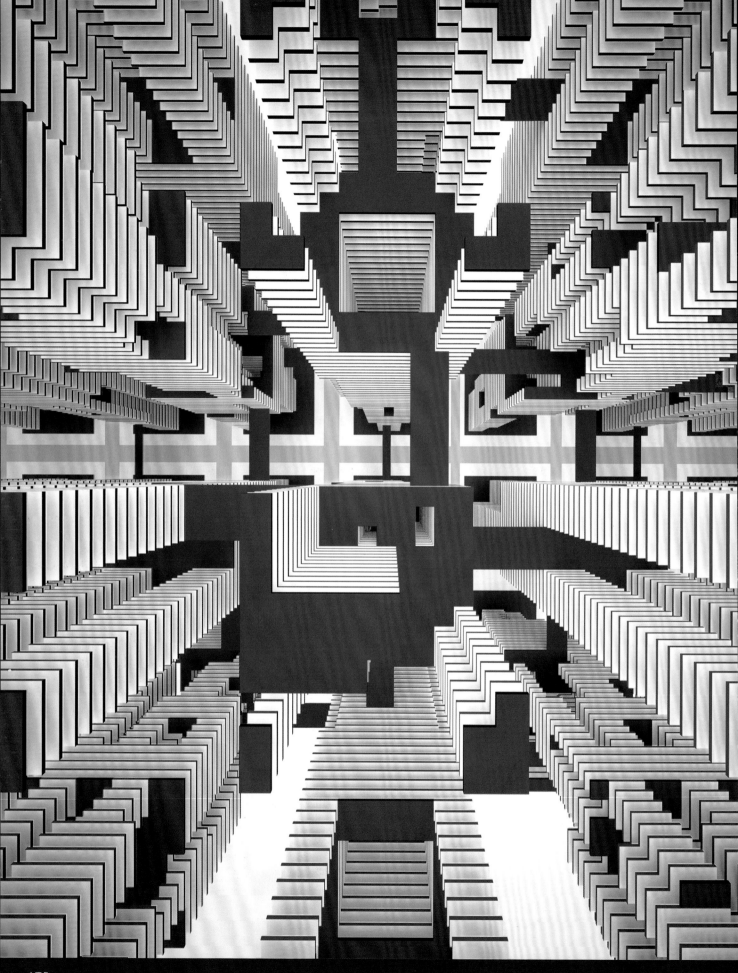

Rocker-Lange Architects, The Ideal City of Refigured Civic Space, 2013

Rocker-Lange's work focuses on subverting standardised systems with computational control, to achieve a credible degree of variation within an urban scale of deployment while pursuing the aim of material realisation of their designs with current technological production systems.

the early 20th century the car was a rarefied luxury item accessible only to the relatively privileged few. What will guarantee that the application of new and emerging technologies will have widespread, pervasive effects on society? The paradigm of mass customisation offers clues as to how to target non-standard design while responding to the contingencies of rapid and large-scale urbanisation. But this view is again biased towards regions such as East Asia that are experiencing the most widespread extent and scale of urbanisation.

Housing the Masses

> So long as the tools a paradigm supplies continue to prove capable of solving the problems it defines, science moves fastest and penetrates most deeply through confident employment of these tools. The reason is clear. As in manufacture so in science – retooling is an extravagance to be reserved for the occasion that demands it. The significance of crises is the indication they provide that an occasion for retooling has arrived.
> — Thomas Kuhn, 'Scientific Revolutions', in *The Philosophy of Science*, 1991, p 78[5]

Industrial production was applied to mass rebuilding after each of the two World Wars, an initiative driven by the need to house the masses. If retooling is reserved for moments of crisis, we may ask: What crisis are we currently experiencing? Twentieth-century housing was not defined so much by elitist Modernist one-off houses commissioned by the wealthy and cultured, but rather by the unrelenting repetitive towers of social and subsidised housing across the world.

In East Asia, social housing has been deemed a great success, most notably in Hong Kong and Singapore. In turn, these repetitive architectural and urban models, which recall the ghost of Le Corbusier's Plan Voisin (1925), are being repeated all over Asia. The migration from rural China to the cities over the last 30 years suggests that if there is a crisis, it may be one of the scale of the problem of housing 10 million people per year, over a 30-year building programme. Standardisation still stands as both the problem of and the solution to the inequities of the world, in particular the poor social and economic status of hundreds of millions of Chinese villagers in new, higher-density and, most often, high-rise urban housing. If not holistically, but rather more piecemeal, to what extent will robotics create opportunities for its application alongside other tools with the capacities to install heterogeneity and distinctiveness, rather than homogeneity and uniformity? The potential for prefabrication (with or without robots) may need to negotiate the limits of where computer numerical control ends and manual construction begins.

If indeed a 'robotic urbanism' will proliferate in the future, will it trickle down from the precision engineering emerging in markets such as Japan, or bottom up from research labs, elite educational institutions or business startups?

If indeed robotics will irreversibly affect the construction industry, as claimed by the designers collated in this issue of △, in what ways will the city of tomorrow be influenced, and how long will it be before its impact takes effect? If indeed a 'robotic urbanism' will proliferate in the future, will it trickle down from the precision engineering emerging in markets such as Japan, or bottom up from research labs, elite educational institutions or business startups? Whichever trajectory the future of construction takes, the driving force of urbanisation is, again, the apparent need to house hundreds of millions of people; this time it is in Asia, and the future of robotics will no doubt be contingent on urbanisation in the Asian Century. △

Notes
1. Tom Verebes, 'The Adaptive City: Urban Change, Resilience, and the Trajectory Towards a Distinctive Urbanism', *DADA: Digital Architectural Design Association, Volume 1*, Tsinghua University Press (Beijing), 2013, p 104.
2. Reyner Banham, *Theory and Design in the First Machine Age*, Butterworth-Heinemann (Oxford), 1960, p 76.
3. Tom Verebes, op cit.
4. Friedrich Kittler, Introduction to 'The History of Communication Media', *CTHEORY.net*, 1996: www.ctheory.net/articles.aspx?id=45.
5. Thomas Kuhn, 'Scientific Revolutions', in Richard Boyd, Philip Gasper and JD Troute (eds), *The Philosophy of Science*, MIT Press (Cambridge, MA), 1991, p 78.

Text © 2014 John Wiley & Sons Ltd. Images: p 126 © Tom Verebes; p 127 © Hanna-Barbera. Everett Collection/REX; p 128 © Qilai Shen/In Pictures/Corbis; pp 130-1 © OCEAN UK Design Ltd (trading as OCEAN CN); p 132 © Rocker-Lange Architects

CONTRIBUTORS

Ralph Bärtschi is a physicist and mathematician with many years' experience in software development and the direct control of computer-controlled machines for fabrication. He has profound knowledge in digital signal processing, geometric algorithms and numerical methodologies. Prior to co-founding ROB Technologies in 2010, he was a senior researcher at the Chair of Architecture and Digital Fabrication, ETH Zurich, where he was primarily responsible for developing control software for diverse research projects on robotic fabrication in architecture.

Thomas Bock's research activities concentrate on the rationalisation of the building industry through automation to reduce costs, and on robotisation to foster mass-customisation – from the planning of building production to modification and deconstruction. He has contributed to the development of five flexible prefabrication systems, 50 construction robots and 10 automated construction sites. He has contributed to around 350 publications. He is a founding director of the International Association for Automation and Robotics in Construction (IAARC), has led a research laboratory for building robotics since 1989, and a human ambient technology laboratory since 1998, at the Technical University of Munich.

Tobias Bonwetsch is cofounder of ROB Technologies, a spinoff company of ETH Zurich that provides software solutions for highly flexible digital fabrication processes. He is also a senior researcher at the Chair of Architecture and Digital Fabrication, ETH Zurich. He studied architecture and graduated from the Technical University of Darmstadt. After gaining practical experience as an independent architect, he completed his postgraduate studies at the ETH Zurich with a specialisation in computer- aided architectural design. His research is concerned with integrating the logic of digital fabrication into the architectural design process, with a special focus on robotic assembly.

Michael Budig studied architecture at the University of Innsbruck, the University of Texas at Arlington and the Bartlett School of Architecture, University College London (UCL). He is a senior researcher at the Singapore-ETH Centre for Global Environmental Sustainability (SEC) and has led the Future Cities Laboratory (FCL) research module for Architecture and Digital Fabrication since 2011. Prior to that he taught at Studio Zaha Hadid at the University of Applied Arts Vienna between 2010 and 2011, and at the Institute for Experimental Architecture, University of Innsbruck, between 2001 and 2011. He is registered as an architect in Austria, and was a principal at moll budig architecture.

Gregory Epps is founder of RoboFold and one of the first people to appropriate robotics for applications in design and architecture. He graduated from the Innovation Design Engineering joint master's at the Royal College of Art and Imperial College London. He is co-organiser of the 'Shape To Fabrication' conference and workshops, and was previously Chair of the Workshops for Advances in Architectural Geometry. He has taught at over 100 workshops and regularly lectures internationally.

Jelle Feringa is a cofounder of EZCT Architecture and Design Research. The work of EZCT is part of the permanent collection of the Centre Pompidou, and the FRAC Centre, Orléans collection, with recent work exhibited at Archilab 2013. While developing his PhD thesis at Hyperbody TU Delft in 2011 he established a robotics lab in Rotterdam harbour. He is Chief Technology Officer at Odico Formwork Robotics and developer of the offline robotics programming platform PyRAPID, which lies at the heart of Odico's operation.

Andreas Froech trained as an architect in Vienna and at Columbia University in New York. From 1997 to 1999 he worked with Greg Lynn and taught at the UCLA School of Architecture in Los Angeles. He was also previously Director of Material Development at Panelite. He has spoken on panels and conferences throughout the US and internationally about his approach to integrated CNC equipment strategy, and has supported as a sponsor many academic projects representing cutting-edge research in the architecture field.

Norman Hack received his Diploma in Architecture from the Technical University of Vienna. A scholarship from the German Academic Exchange Organization (DAAD) allowed him to pursue a postgraduate degree at the Architectural Association (AA) in London, from which he graduated with distinction. He gained professional experience in renowned offices across Europe, including Coop Himmelb(l)au, UNStudio and Herzog & de Meuron where he worked as a specialist in computational design and fabrication. His PhD research at the Chair of Architecture and Digital Fabrication at ETH Zurich focuses on material processes for constructive non-standard assemblies.

Stephan Henrich studied architecture and urban design at the University of Stuttgart and the École d´Architecture de Paris Belleville. From 2003 to 2006 he worked for Knippers Helbig Advanced Engineering in Stuttgart, and in 2005 was involved with the New-Territories/R&Sie(n) research/exhibition project 'I´ve heard about' that was shown at the Musée d'Art Moderne (MAM), Paris in 2009, and 'Une Architecture Des Humeurs' by R&Sie(n) shown at Le Laboratoire, Paris; the Tinguely Museum, Basle; the Kunsthaus Graz; and the Venice Architectural Biennale.

Volker Helm completed his studies in architecture at the University of Siegen (Germany), before specialising in CAAD as a Master of Advanced Studies under the Chair of Professor Ludger Hovestadt at ETH Zurich. His studies focused on computer-aided architectural design and automated production. He also worked for six years at Herzog & de Meuron in Basle, with a focus on the development, programming and realisation of complex geometries. Since 2010 he has been researching at the Chair of Architecture and Digital Fabrication at ETH Zurich. His doctoral thesis focuses on robot-based construction processes on site.

Ryan Luke Johns is a research specialist at the Princeton University School of Architecture, New York, and cofounder of GREYSHED, a design-research collaborative focusing on robotic fabrication within architecture, art and industrial design. He holds a BA in Architecture with a concentration in mathematics from Columbia University, and a Master of Architecture from Princeton University.

Willi Viktor Lauer is a research assistant at the Future Cities Laboratory (FCL) at the Singapore-ETH Centre for Global Environmental Sustainability (SEC) where he implemented a research facility for investigating robotic fabrication methods for high rises. Between 2009 and 2011 he worked as a scientific assistant at the Chair of Building Realisation and Robotics at the Technical University Munich, where he gained deep insights into the young history of robotic construction technologies and the forerunning building industrialisation. In the context of his Master's thesis in 2009, he reconstructed the first architectural robotic arm – the Location Orientation Manipulator by Konrad Wachsmann.

Camille Lacadée is a cofounder of [eIf/bʌt/c] institute for contingent scenario (film and architecture) with François Roche.

She graduated in 2009 from the École Spéciale d'Architecture in Paris, after passing her RIBA Part 1 in 2008 at the AA in London. Since then she has lived and worked in Asia (Japan, India and Thailand) and is currently co-leading the New-Territories [eIf/bʌt/c] architecture studio in Bangkok while pursuing the design of a cultural centre in Karnataka, India.

Silke Langenberg is Professor for Design and Construction in Existing Contexts, Conservation and Building Research at the University of Applied Sciences in Munich. She was previously a senior researcher at the Institute of Technology in Architecture (Chair of Architecture and Digital Fabrication, 2012–13) and at the Institute for Conservation and Building Research (2006–11) at ETH Zurich. Her research focuses on the attempts to optimise planning and to rationalise the building process during the 1960s and 1970s, as well as on questions concerning the development, repair and conservation of system buildings.

Jason Lim graduated with a Bachelor of Architecture from Cornell University, New York, and was awarded the AIA Certificate of Merit. He subsequently studied at the Product Architecture Lab at the Stevens Institute of Technology, and received a Master of Engineering degree. He has worked at BriggsKnowles Architecture + Design in New York City, and also taught at the School of Constructed Environments at Parsons The New School for Design, and at the Art and Technology programme at the Stevens Institute. Since September 2011 he has been a research assistant at the Future Cities Laboratory (FCL) at the Singapore-ETH Centre for Global Environmental Sustainability (SEC), where he is a PhD student of Fabio Gramazio and Matthias Kohler.

Philippe Morel is an architect and theorist, and cofounder of EZCT Architecture & Design Research (2000). He is an associate professor at the ENSA Paris-Malaquais where he directs the Digital Knowledge programme, as well as an invited research cluster master at The Bartlett School of Architecture, UCL. Prior to The Bartlett, he taught at the Berlage Institute (seminar and studio) and the AA (HTS Seminar and AADRL studio). His long-standing interest in the elaboration of a theory of computational architecture is expressed by some of his first published essays. He has lectured in various places, including the MIT Department of Architecture and the AA. He is also author of the book *Empiricism & Objectivity: Architectural Investigations with Mathematica* (2003–4), subtitled *A Coded Theory for Computational Architecture*, which is written entirely in code.

Neri Oxman directs the Mediated Matter group at the MIT Media Lab, which focuses on biologically inspired design fabrication tools and technologies aiming to enhance the relationship between natural and man-made environments. The group's research field integrates computational form-finding strategies with digital fabrication. The goal is to establish new forms of design and novel processes of material practice at the intersection of computer science, material engineering, design and ecology, with broad applications across multiple scales.

Raffael Petrovic studied architecture with Studio Zaha Hadid at the University for Applied Arts Vienna. After an exchange semester at UCLA and a semester project at Crossover Studio François Roche, he graduated in 2011. During his studies he focused on parametric design techniques and rapid prototyping technologies, and their potential to generate new architectural typologies. Since July 2012 he has been a teaching and research assistant at the Future Cities Laboratory (FCL) at the Singapore-ETH Centre for Global Environmental Sustainability (SEC).

Antoine Picon is the G Ware Travelstead Professor of the History of Architecture and Technology at Harvard Graduate School of Design (GSD) where he also co-chairs the doctoral programmes. He has published numerous books and articles mostly dealing with the complementary histories of architecture and technology. His *Digital Culture in Architecture* (Birkhäuser, 2010) proposes a comprehensive interpretation of the changes brought by the computer to the design professions. His most recent book, *Ornament: The Politics of Architecture and Subjectivity* (John Wiley & Sons, 2013), deals with the relationship between digital culture and the 'return' of ornament in architecture.

François Roche is the principal of New-Territories (R&Sie(n)/[eIf/bʌt/c]). He is based mainly in Bangkok [eIf/bʌt/c], sometimes in Paris (R&Sie(n)), and spends autumn in New York with his research studio at the Graduate School of Architecture, Preservation and Planning (GSAPP) at Columbia University. Through these different structures, his architectural works and protocols seek to articulate the real and/or fictional, the geographic situations and narrative structures that can transform them.

Asbjørn Søndergaard is Chief Development Officer and a founding partner in Odico Formwork Robotics. As an architectural researcher and coordinator of Digital Experimentation at the Aarhus School of Architecture, he works in the field of digital fabrication in its relation to architectural design. His ongoing doctoral research focuses on topological optimisation of architectural structures and the development of novel structural logics in relation to robotic fabrication techniques.

Tom Verebes is the founder and Creative Director of OCEAN CN Consultancy Network, based in Hong Kong, with collaborators in Beijing, Shanghai and London. He is currently Associate Professor in the Department of Architecture at the University of Hong Kong (HKU), and Director of the AA Shanghai Visiting School. Past positions include Associate Dean (Teaching & Learning) in the Faculty of Architecture at HKU (2011–14); Co-Director of the AA Design Research Lab (DRL) at the Architectural Association in London, where he taught design studio and seminars in the post-professional MArch (1997–2009); and Guest/Visiting Professor at ABK Stuttgart (2004–6). He has written, published, exhibited and lectured extensively in Europe, North America, Asia and the Middle East.

Jan Willmann is senior assistant at the Chair of Architecture and Digital Fabrication at ETH Zurich. He studied architecture in Liechtenstein, Oxford and Innsbruck where he received his PhD degree in 2010. He was previously a research assistant and lecturer at the Chair of Architectural Theory of Professor Ir Bart Lootsma and gained professional experience in numerous architectural offices. His research focuses on digital architecture and its theoretical implications as a composed computational and material score. He has lectured and exhibited internationally, and published extensively, including *The Robotic Touch: How Robots Change Architecture* (Park Books, 2014) together with Fabio Gramazio and Matthias Kohler.

ABOUT ARCHITECTURAL DESIGN

INDIVIDUAL BACKLIST ISSUES OF △ ARE AVAILABLE FOR PURCHASE AT £24.99 / US$45

TO ORDER AND SUBSCRIBE SEE BELOW

What is *Architectural Design*?

Founded in 1930, *Architectural Design* (△) is an influential and prestigious publication. It combines the currency and topicality of a newsstand journal with the rigour and production qualities of a book. With an almost unrivalled reputation worldwide, it is consistently at the forefront of cultural thought and design.

Each title of △ is edited by an invited guest-editor, who is an international expert in the field. Renowned for being at the leading edge of design and new technologies, △ also covers themes as diverse as architectural history, the environment, interior design, landscape architecture and urban design.

Provocative and inspirational, △ inspires theoretical, creative and technological advances. It questions the outcome of technical innovations as well as the far-reaching social, cultural and environmental challenges that present themselves today.

For further information on △, subscriptions and purchasing single issues see: www.architectural-design-magazine.com

How to Subscribe

With 6 issues a year, you can subscribe to △ (either print, online or through the △ App for iPad).

INSTITUTIONAL SUBSCRIPTION
£212/US$398 print or online

INSTITUTIONAL SUBSCRIPTION
£244/US$457 combined print & online

PERSONAL-RATE SUBSCRIPTION
£120/US$189 print and iPad access

STUDENT-RATE SUBSCRIPTION
£75/US$117 print only

To subscribe to print or online:
Tel: +44 (0) 1243 843 272
Email: cs-journals@wiley.com

△ APP FOR iPAD
For information on the △ App for iPad go to www.architectural-design-magazine.com
6-issue subscription: £44.99/US$64.99
Individual issue: £9.99/US$13.99

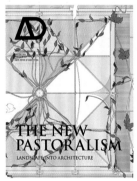

Volume 83 No 3
ISBN 978 1118 336984

Volume 83 No 4
ISBN 978 1118 361429

Volume 83 No 5
ISBN 978 1118 418796

Volume 83 No 6
ISBN 978 1118 361795

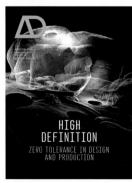

Volume 84 No 1
ISBN 978 1118 451854

Volume 84 No 2
ISBN 978 1118 452721